Differential Equations with MATLAB®

(Second Edition)

Brian R. Hunt
Ronald L. Lipsman
John E. Osborn
Jonathan M. Rosenberg

Differential Equations with MATLAB®

(Second Edition)

Updated for MATLAB® 7 and Simulink® 6

Brian R. Hunt
Ronald L. Lipsman
John E. Osborn
Jonathan M. Rosenberg
All of the University of Maryland, College Park

with

Kevin R. Coombes
Garrett J. Stuck

JOHN WILEY & SONS, INC.

MATLAB, Simulink, and Handle Graphics are registered trademarks of the MathWorks, Inc.

To order books or for customer service, please call 1-800-CALL-WILEY (225-5945).

ISBN 0-471-71812-2

10 9 8 7 6 5 4 3 2

Preface

As the subject matter of differential equations continues to grow, as new technologies become commonplace, as old areas of application are expanded, and as new ones appear on the horizon, the content and viewpoint of courses and their textbooks must also evolve.

Boyce & DiPrima, **Elementary Differential Equations**, Eighth Edition.

Traditional introductory courses in ordinary differential equations (ODE) have concentrated on teaching a repertoire of techniques for finding formula solutions of various classes of differential equations. Typically, the result was rote application of formula techniques without a serious qualitative understanding of such fundamental aspects of the subject as stability, asymptotics, dependence on parameters, and numerical methods. These fundamental ideas are difficult to teach because they have a great deal of geometrical content and, especially in the case of numerical methods, involve a great deal of computation. Modern mathematical software systems, which are particularly effective for geometrical and numerical analysis, can help to overcome these difficulties. This book changes the emphasis in the traditional ODE course by using a mathematical software system to introduce numerical methods, geometric interpretation, symbolic computation, and qualitative analysis into the course in a basic way.

The mathematical software system we use is MATLAB®. (This book is also available in *Mathematica* and *Maple* versions.) We assume that the user has no prior experience with MATLAB. We include concise instructions for using MATLAB on three popular computer platforms: *Windows*, *Macintosh*, and UNIX workstations, including *Linux*. This book is not a comprehensive introduction or reference manual to either MATLAB or any of the computer platforms. Instead, it focuses on the specific features of MATLAB that are useful for analyzing differential equations. A new addition in this second edition is a discussion of Simulink®, an auxiliary to MATLAB that is increasingly popular among scientists and engineers as a simulation tool. Again, we focus on the aspects of Simulink most useful for numerical and graphical solution of differential equations.

This supplement can easily be used in conjunction with most ODE textbooks. It addresses the standard topics in ordinary differential equations, but with a substantially different emphasis.

We had two basic goals in mind when we introduced this supplement into our course. First, we wanted to deepen students' understanding of differential equations by giving them a new tool, a mathematical software system, for analyzing differential equations. Second, we wanted to bring students to a level of expertise in the mathematical software system that would allow them to use it in other mathematics, engineering, or science courses. We believe that we have achieved these goals in our own classes. We hope this supplement will be useful to students and instructors on other campuses in achieving the same goals.

Acknowledgment and Disclaimer. We are pleased to acknowledge support of our research by the National Science Foundation, which contributed over many years to the writing of this book. Our work on the second edition was partially supported by NSF Grants DMS-0103647, DMS-0104087, and DMS-0341982. Any opinions, findings, and conclusions or recommendations expressed in this material are those of the authors and do not necessarily reflect the views of the National Science Foundation.

Brian R. Hunt
Ronald L. Lipsman
John E. Osborn
Jonathan M. Rosenberg

College Park, Maryland
October, 2004

Contents

Chapter 1

Introduction

We begin by describing the philosophy behind our approach to the study of ordinary differential equations. This philosophy has its roots in the way we understand and apply differential equations; it has influenced our teaching and guided the development of this book. This chapter also contains two user's guides, one for students and one for instructors.

1.1 Guiding Philosophy

In scientific inquiry, when we are interested in understanding, describing, or predicting some complex phenomenon, we use the technique of mathematical modeling. In this approach, we describe the state of a physical, biological, economic, or other system by one or more functions of one or several variables. For example:

- The position $s(t)$ of a particle is a function of the time t.
- The temperature $T(x, y, z, t)$ in a body is a function of the position (x, y, z) and the time t.
- The populations $x(t)$ and $y(t)$ of two competing species are functions of time.
- The gravitational or electromagnetic force on an object is a function of its position.
- The money supply is a function of time.

Next, we attempt to formulate, in mathematical terms, the fundamental law governing the phenomenon. Typically, this formulation results in one or more differential equations; *i.e.*, equations involving derivatives of the functions describing the state of the system with respect to the variables they depend on. Frequently, the functions depend on only one variable, and the differential equation is called *ordinary*. To be specific, if $x(t)$ denotes the function describing the state of our system, an *ordinary differential equation* for $x(t)$ might involve $x'(t)$, $x''(t)$, higher derivatives, or other known functions of t. By contrast, in a *partial differential equation*, the functions depend on several variables.

In addition to the fundamental law, we usually describe the initial state of the system. We express this state mathematically by specifying *initial values* $x(0)$, $x'(0)$, *etc.* In this way, we arrive at an *initial value problem*—an ordinary differential equation together with initial conditions. If we can solve the initial value problem, then the solution is a function $x(t)$ that predicts the future state of the system. Using the solution, we can describe qualitative or quantitative properties of the system. At this stage, we compare the values predicted by $x(t)$ against experimental data accumulated by observing the system. If the experimental data and the function values match, we assert our model accurately models the system. If they do not match, we go back, refine the model, and start again. Even if we are satisfied with the model, new technology, new requirements, or newly discovered features of the system may render our old model obsolete. Again, we respond by reexamining the model.

The subject of differential equations consists in large part of building, solving, and analyzing mathematical models. Many results and methods have been developed for this purpose. These results and methods fall within one or more of the following themes:

1. Existence and uniqueness of solutions,
2. Dependence of solutions on initial values,
3. Derivation of formulas for solutions,
4. Numerical calculation of solutions,
5. Graphical analysis of solutions,
6. Qualitative analysis of differential equations and their solutions.

The basic results on existence and uniqueness (theme 1) and dependence on initial values (theme 2) form the foundation of the subject of differential equations. The eminent French mathematician Jacques Hadamard referred to a problem as *well-posed* if good results are available concerning these two themes, *i.e.*, if one has existence and uniqueness of solutions and if small changes in initial values make only a small change in the solution. The derivation of formula solutions (theme 3) is a rich and important part of the subject; a variety of methods have been developed for finding formula solutions to special classes of equations. Although many equations can be solved exactly, many others cannot. However, any equation, solvable or not, can be analyzed using numerical methods (theme 4), graphical methods (theme 5), and qualitative methods (theme 6). The ability to obtain qualitative and quantitative information without the aid of an explicit formula solution is crucial. That information may suffice to analyze and describe the original phenomenon (which led to the model, which gave rise to the differential equation).

Traditionally, introductory courses in differential equations focused on methods for deriving exact solutions to special types of equations and included some simple numerical and qualitative methods. The human limitations involved in compiling numerical or graphical data were formidable obstacles to implementing more advanced qualitative or quantitative methods. Computer platforms have reduced these obstacles. Sophisticated software and mainframe computers enhanced the use of quantitative and qualitative methods in the theory and applications of differential equations. With the arrival of comprehensive mathematical software systems on personal computers, this modern approach has become accessible to students.

In this book, we use the mathematical software system MATLAB to implement this approach. We use MATLAB's symbolic, numerical, and graphical capabilities to analyze differential equations and their solutions.

Finally, engineers and scientists have to develop not only skills in analyzing problems and interpreting solutions, but also the ability to present coherent conclusions in a logical and convincing style. Students should learn how to submit solutions to the computer assignments in such a style. This is excellent preparation for professional requirements that lie ahead.

1.2 Student's Guide

The chapters of this book can be divided into three classes: general discussion of MATLAB, supplementary material on ordinary differential equations (ODE), and computer problem sets. Here is a brief description of the contents.

Chapter 2 explains how to start and run MATLAB on your computer. Chapter 3 introduces basic MATLAB commands. Unless you have previous experience with MATLAB, you should work through Chapter 3 while sitting at your computer. Then you should read Chapter 4, which contains detailed instructions for using MATLAB M-files and printing or "publishing" your work. After that, work the problems in Problem Set A to practice the skills you've learned in Chapters 3 and 4. These steps will bring you to a basic level of competence in the use of MATLAB, sufficient for the first three of the differential equations chapters, Chapters 5–7, and for Problem Set B. Some more advanced aspects of MATLAB, needed for some of the problems starting in Problem Set C, are discussed in Chapter 8. Chapter 9 introduces the MATLAB companion program known as Simulink, which provides a useful graphical tool for solving initial value problems.

Since the primary purpose of this book is to study differential equations, we have not attempted to describe all the major aspects of MATLAB or Simulink. You can explore MATLAB in more depth using its demos, tutorials, and online help, or by consulting more comprehensive books such as:

- Amos Gilat, **MATLAB: An Introduction with Applications**, 2nd edition, John Wiley & Sons, Inc., 2005.
- Brian R. Hunt, Ronald L. Lipsman, and Jonathan M. Rosenberg, **A Guide to MATLAB: for Beginners and Experienced Users**, Cambridge University Press, 2001. (This edition was written for MATLAB 6; a 2nd edition for MATLAB 7 is in preparation.)
- William J. Palm III, **Introduction to MATLAB 7 for Engineers**, McGraw-Hill, 2005.

- Rudra Pratap, **Getting Started with MATLAB: A Quick Introduction for Scientists & Engineers, Version 7**, Oxford University Press, 2005.

- Timothy A. Davis and Kermit Sigmon, **MATLAB Primer**, 7th edition, Chapman & Hall/CRC, 2005.

The eight ODE chapters (5–7, 10–14) are intended to supplement the material in your text. The emphasis in this book differs from that found in a traditional ODE text. The main difference is less emphasis on the search for exact formula solutions, and greater emphasis on qualitative, graphical, and numerical analysis of the equations and their solutions. Furthermore, the commands for analyzing differential equations with MATLAB appear in these chapters.

The six computer problem sets form an integral part of the book. Solving these problems will expose you to the qualitative, graphical, and numerical features of the subject. Each set contains about twenty problems.

You can most profitably attack the problem sets if you plan to do them in two sessions. Begin by reading the problems and thinking about the issues involved. Then go to the computer and start solving the problems. If you get stuck, save your work and go on to the next problem. If there are things you'd like to discuss with your instructor, print out the relevant parts of your input and output. Talk to your instructor or your peers about anything you don't understand. This first session should be attempted well before the assignment is due. After you have reviewed your output and obtained answers to your questions, you are ready for your second session. At this point, you should fill the gaps, correct your mistakes, and polish your M-file. Although you may find yourself spending extra time on the first few problems, if you read the MATLAB chapters carefully, and follow the suggestions above, you should steadily increase your level of competence in using MATLAB.

The end of this book contains two useful sections: a Glossary and a collection of Sample Solutions. The Glossary contains a brief summary of relevant MATLAB commands, built-in functions, and programming constructs. The Sample Solutions show how we solved several problems from this book. These samples can serve as guides when you prepare your own solutions. Emulate them. Strive to prepare coherent, organized solutions. Combine MATLAB's input, output, and graphics with your own textual commentary and analysis of the problem. Edit the final version of your solution to remove syntax errors and false starts. You will soon take pride in submitting complete, polished solutions to the problems.

1.3 Instructor's Guide

The philosophy that guided the writing of this book is explained at the beginning of this chapter. Here is a capsule summary of that philosophy. We seek:

- To guide students into a more interpretive mode of thinking.

- To use a mathematical software package to enhance students' ability to compute symbolic and numerical solutions, and to perform qualitative and graphical analysis of differential equations.

- To develop course material that reflects the current state of ODE and emphasizes the mathematical modeling of physical problems.

- To minimize the time required to learn to use the software package.

As mentioned in Section 1.2, our material consists of MATLAB discussion, ODE supplements, and computer problem sets. Here are our recommendations for integrating this material into a typical first course in differential equations for scientists or engineers.

1.3.1 MATLAB and Simulink

Our students read Chapters 2, 3, and 4, and work Problem Set A within the first week of the semester. Although Chapter 8 is not essential for Problem Set B, students often find it useful. Attention to these chapters quickly leads students to a basic level of proficiency. Chapter 9 teaches enough about Simulink to make it possible for students to use it for many problems in Problem Sets C-F.

1.3.2 ODE Chapters

These eight chapters (5–7 and 10–14) supplement the material in a traditional text. We use MATLAB to study differential equations using symbolic, numerical, graphical, and qualitative methods. We emphasize the following topics: direction fields, stability, numerical methods, comparison methods, and phase portraits. These topics are not emphasized to the same degree in traditional texts. We incorporate this new emphasis into our class discussions, devoting some class time to each chapter. Specific guidelines are difficult to prescribe, and the required time varies with each chapter, but on average we spend up to an hour per chapter in class discussion.

The structure of this book requires that numerical methods be discussed early in the course, immediately after the discussion of first order equations. The discussion of numerical methods is directed toward the use of **`ode45`**, MATLAB's primary numerical ODE solver.

1.3.3 Computer Problem Sets

There are six computer problem sets. The topics addressed in the problem sets are:

(A) Practice with MATLAB,

(B) First Order Equations,

(C) Numerical Solutions,

(D) Second Order Equations,

(E) Series Solutions and Laplace Transforms,

(F) Systems of Differential Equations.

Problem Set A is a practice set designed to acquaint students with the basic symbolic and graphical capabilities of MATLAB, and to reacclimate them to calculus. We assign all problems in Problem Set A, and have it turned in rather quickly. We generally assign 3–5 problems from each of the remaining problem sets.

In addition to analyzing problems critically, it is important that students present their analyses in coherent English and mathematics, displayed appropriately on their printouts. To accomplish this, they must master M-files, and the "publish" feature of MATLAB (the **`publish`** command), M-books, or the pasting of MATLAB code and output into word processing documents. These tools facilitate presentations with integrated input, output, graphics, and formatted text. Chapter 4 contains detailed instructions on the "publish" feature and on M-books. Engineers and scientists do not just solve problems; they must also present their ideas in a cogent and convincing fashion. We expect students to do the same in our course. To encourage students to submit high-quality solutions to the homework problems, we have provided *Sample Solutions* to selected problems at the end of this book. These solutions were prepared using MATLAB's **`publish`** command.

We should mention that we use the text **Elementary Differential Equations**, eighth edition, by William E. Boyce & Richard C. DiPrima, John Wiley & Sons, Inc., 2005, in our course. The references to Boyce & DiPrima in our book are all to this edition. We have found that our book is easily integrated with this text. We believe our book can likewise be conveniently integrated with any other text for a first course in differential equations for scientists or engineers. Some suggestions on how to accomplish this may be found on our web site:

`http://www.math.umd.edu/undergraduate/schol/ode`

1.4 A Word About Software Versions

New versions of software appear frequently. When a complex program like MATLAB changes, many commands work better than they did before, some work differently, and a few may no longer work at all. As this book goes to press, the current version is MATLAB Version 7 (Release 14); this is the version we describe.

Chapter 2

Getting Started with MATLAB

In this chapter, we will introduce you to the tools you need in order to begin using MATLAB effectively. These include: some relevant information on computer platforms and software; installation protocols; how to launch MATLAB, enter commands, use online help, and recover from hang-ups; a roster of MATLAB's various windows; and finally, how to exit the program. We know you are anxious to get started using MATLAB, so we will keep this chapter brief. After you complete it, you can go immediately to Chapter 3 to find concrete and simple instructions for using MATLAB to do mathematics. We describe the MATLAB interface more elaborately in Chapter 4, and we start in earnest on differential equations in Chapter 5.

2.1 Platforms and Versions

It is likely that you will run MATLAB on a PC (running Windows or Linux), or on some form of UNIX operating system. Some previous versions of MATLAB (Releases 11 and 12) did not support Macintosh, but the most current versions (Releases 13 and 14) do. (On a Macintosh, MATLAB 7 requires Mac OS X 10.3.2 (Panther).) If you are running a Macintosh platform, you should find that our instructions for Windows platforms will suffice for your needs. Like MATLAB 6 (Releases 12 and 13), and unlike earlier versions, MATLAB 7 looks virtually identical on Windows and UNIX platforms. For definitiveness, we shall assume the reader is using a PC in a Windows environment. In those very few instances where our instructions must be tailored differently for Linux, UNIX or Macintosh users, we shall point it out clearly.

Remark 2.1 We use the word Windows to refer to all flavors of the Windows operating system. MATLAB 7 (R14) *will* run on Windows 2000, Windows NT (4.0 and higher), and on Windows XP. MATLAB 7 will *not* run on Windows 95, Windows 98, or on Windows ME. However, MATLAB 6.5 (R13) *does* run on Windows 98 or Windows ME.

This book is written to be compatible with the current version of MATLAB, namely MATLAB 7 (R14). The vast majority of the MATLAB commands we describe, as well as many features of the MATLAB interface (*e.g.*, M-files and M-books) are valid for version 6.5 (R13), and earlier versions in some cases. We also note that the differences between the MATLAB Professional Version and the MATLAB Student Version are rather minor, and virtually unnoticeable to a beginner, or even a mid-level user. Again, in the few instances where we describe a MATLAB feature that is only available in the Professional Version, we highlight that fact clearly.

2.2 Installation

If you intend to run MATLAB on a PC, especially the Student Version, it is quite possible that you will have to install it yourself. You can easily accomplish this using the product CDs. Follow the installation instructions as you would with any new software installation. At some point in the installation you may be asked which *toolboxes* you wish to install. Unless you have severe space limitations, we suggest that you install any that seem of interest to you or that you think you might use at some point in the future. However, for the purposes of this course, you should be sure to include the *Symbolic Math Toolbox*. We also strongly encourage you to install Simulink, which is described in Chapter 9.

2.3 Starting MATLAB

You start MATLAB as you would any other software application. On a PC you access it via the **Start** menu, in the **Programs** folder under **MATLAB 7.0** or **Student MATLAB**. Alternatively, you may have a desktop icon that enables you to start MATLAB with a simple double-click. On a UNIX machine, generally you need only type **`matlab`** in a terminal window, though you may first have to find the `matlab7.0/bin` directory and add it to your path. Or you may have an icon or a special button on your desktop that achieves the task.

However you start MATLAB, you will briefly see a window that displays the MATLAB logo as well as some MATLAB product information, and then a *MATLAB Desktop* window will launch. That window will contain a title bar, a menu bar, a tool bar and four embedded windows, one of which is hidden. The largest and most important window is the *Command Window* on the right. We will go into more detail in Chapter 4 on the use and manipulation of the other three windows: the *Workspace Browser*, the *Command History Window*, and the *Current Directory Browser*. For now we concentrate on the Command Window in order to get you started issuing MATLAB commands as quickly as possible. At the top of the Command Window, you may see some general information about MATLAB, perhaps some special instructions for getting started or accessing help, but most important of all, you will see a command prompt (>> or EDU >>). If the Command Window is "active", its title bar will be dark, and the prompt will be followed by a cursor (a blinking vertical line). That is the place where you will enter your MATLAB commands (see Chapter 3). If the Command Window is not active, just click in it anywhere. Figure 2.1 contains an example of a newly launched MATLAB Desktop.

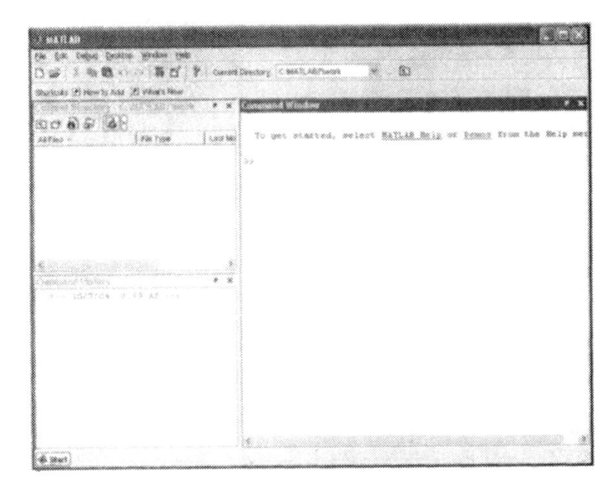

Figure 2.1: A MATLAB Desktop

Remark 2.2 MATLAB 6 has a Desktop, but in older versions of MATLAB, for example 5.3, there was no integrated Desktop. Only the Command Window appeared when you launched the application. (On UNIX systems, the terminal window from which you invoked MATLAB 5 became the Command Window.) Commands that we instruct you to enter in the Command Window inside the Desktop for version 7 can be entered directly into the Command Window in any older version.

2.4 Typing in the Command Window

Click in the Command Window to make it active. When a window becomes active, its titlebar darkens and a blinking cursor appear after the prompt. Now you can begin entering commands. Try typing **`2+2`**, then press the ENTER or RETURN key. Next try **`factor(987654321)`**. And finally **`cos(100)`**. Your MATLAB Desktop should look like Figure 2.2.

2.5 Online Help

MATLAB has extensive online help. In fact, using only this book and the online help, you should be able to become quite proficient with MATLAB.

You can access the online help in one of several ways. Typing **`help`** at the command prompt will reveal a long list of topics for which help is available. Just to illustrate, try typing **`help general`**. Now you see a long list of "general purpose" MATLAB commands. Finally, try **`help solve`** to learn about the **`solve`** command. In every instance above, more information than your screen can hold will scroll by. You can use the scroll bar on

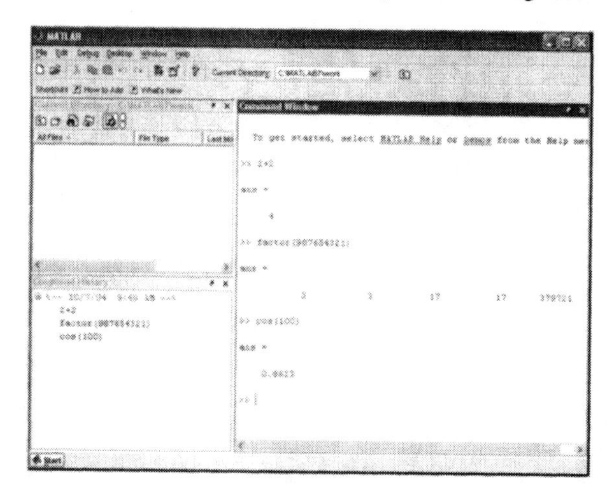

Figure 2.2: The MATLAB Desktop with Several Commands Evaluated

the right of the window to scroll back up. Alternatively, you can force MATLAB to display information one screenful at a time by typing **`more on`**. You press the space bar to display the next screenful; type **`help more`** for details. Typing **`more on`** affects all subsequent commands, until you type **`more off`**.

The **`lookfor`** command searches the first line of every MATLAB help file for a specified string (use **`lookfor -all`** to search all lines). For example, if you wanted to see a list of all MATLAB commands that contain the word "factor" as part of the command name or brief description, then you would type **`lookfor factor`**. If the command you are looking for appears in the list, then you can use **`help`** on that command to learn more about it.

There is a much more user friendly way to access the online help, namely via the MATLAB *Help Browser*. You can activate it in several ways, for example, typing **`helpwin`**, **`helpdesk`** or **`doc`** at the command prompt brings it up. Alternatively, it is available through the menu bar under **Help**. Finally, the question mark button on the tool bar will also invoke the Help Browser. Upon its launch you will see two windows. The first is called the *Help Navigator*; it is used to find documentation. The second, called the *display pane*, is used for viewing documentation. The display pane works much like a normal web browser. It has an address window, buttons for moving forward and back (among the windows you have visited), hyperlinks for moving around in the documentation, the capability of storing favorite pages, and other useful tools.

A particularly useful way to invoke the Help Browser is to type **`doc solve`**. This launches the Help Browser and displays the reference page for **`solve`**. The reference page for a command is generally similar to the text available (through **`help`**), but sometimes has more information.

You can also use the Help Navigator to locate the documentation that you will explore in the display pane. The Help Navigator has four tabs that allow you to arrange your search for documentation in different ways. The first is the *Contents* tab that displays a tree view of all the documentation topics available. The extent of that tree will be determined by how much you (or your system administrator) included in the original MATLAB installation (how many toolboxes, *etc.*). The second tab is an *Index* that displays all available documentation in index format. It responds to your key entry of likely items you want to investigate in the usual alphabetic reaction mode. The third tab provides the *Search* mechanism. You type in what you seek, either a function or some other descriptive term, and the search engine locates documentation pertaining to your entry. Finally, the fourth tab is a small collection of *Demos* that you can run (in various media) to learn more about MATLAB. Clicking on an item that appears in any of these tabs brings up the corresponding documentation in the display pane.

The Help Browser has an excellent tutorial describing its own operation. To view it, open the browser; if the display pane is not displaying the "Begin Here" page, then click on it in the Contents tab; scroll down to the "Getting Help" link in the display pane and click on it; or reopen the Help browser by typing **doc** in the command window. The Help Browser is a powerful and easy-to-use aid in finding the information you need on various features of MATLAB. Like any such tool, the more you use it, the more adept you become at its use.

Remark 2.3 If you type **helpwin** to launch the Help Browser, the display pane will contain the same roster that you see as the result of typing **help** at the command prompt, but the entries will be links.

To summarize, the Help Browser is a robust hypertext browser that you can use, by clicking, to browse through a host of command and interface information. Figure 2.3 depicts the *MATLAB Help Browser*.

Remark 2.4 If you are working with MATLAB version 5.3 or earlier, then typing **help**, **help general** or **help solve** at the command prompt will work as indicated above. But the entries **helpwin** or **helpdesk** call up more primitive, although still quite useful, forms of help windows than the robust *Help Browser* available with version 7.

If you are patient, and not overly anxious to get to Chapter 3, you can type **demo** to try some of MATLAB's online demonstrations.

2.6 MATLAB Windows

We have already described the MATLAB Command Window and the Help Browser, and have mentioned in passing the Command History Window, the Current Directory Browser, and the Workspace Browser. These, and several other windows you will encounter as you work with MATLAB, will allow you to: control files and folders that you and MATLAB will need to access; write and edit the small MATLAB programs (M-files) that you will use to run MATLAB effectively; keep track of the variables and functions that you define as

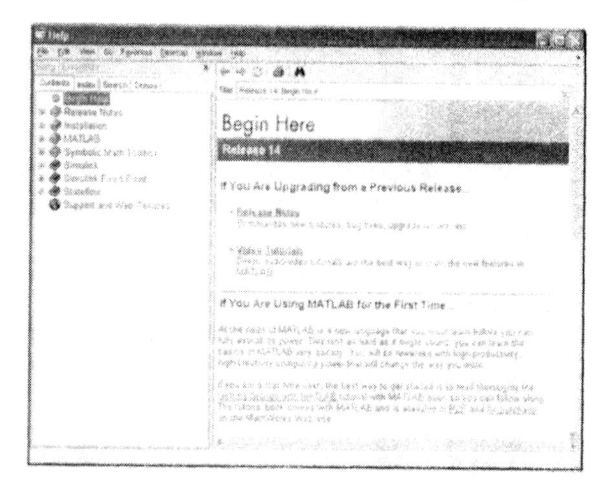

Figure 2.3: The MATLAB Help Browser

you use MATLAB; and design graphical models to solve problems and simulate processes. Some of these windows launch separately, and some are embedded in the Desktop. You can dock those that launch separately inside the Desktop through the **Desktop ▶ Dock** menu button, or by clicking on the downward curved arrow in the toolbar. You can separate windows inside your MATLAB Desktop out to your computer desktop by clicking on the upward curved arrow in the upper right corner of the window's toolbar.

These features are described more thoroughly in Chapters 4 and 8. For now, we want to call your attention to the other main type of window you will encounter, namely graphics windows. Many of the commands you issue will generate graphics or pictures. These will appear in a separate window. MATLAB documentation refers to these as *figure windows*. In this book, we shall also call them graphics windows. In Chapter 8, we will teach you how to generate and manipulate MATLAB graphics windows effectively. See Figure 3.1 in Chapter 3 for a simple example of a graphics window.

Remark 2.5 In MATLAB 6 or earlier versions, you cannot dock figure windows. Nor can you in version 7 if you are operating on a Macintosh platform.

2.7 Ending a Session

The simplest way to conclude a MATLAB session is to type **`quit`** at the prompt. You can also click on the special symbol that closes your windows (usually an × in the upper right hand corner). Still another way to exit is to use the **Exit MATLAB** option from the **File** menu of the Desktop. *Before* you exit MATLAB, you should be sure to save your work, print any graphics or other files you need, and in general clean up after yourself. Some strategies for doing so are addressed in Chapter 4.

Chapter 3

Doing Mathematics with MATLAB

This chapter describes some of the basic MATLAB commands that you will use in this book. We recommend that you start MATLAB and enter the commands displayed in this chapter as you read it.

3.1 Arithmetic

MATLAB can do arithmetic just like a calculator. You can add with **+**, subtract with **-**, multiply with *****, divide with **/**, and exponentiate with **^**. To evaluate an arithmetic expression, type the expression at the **>>** prompt and then press the ENTER key or RETURN key. For example:

```
>> 2^3 - (4 + 5)/6 + 7*8
ans =
   62.5000
```

MATLAB prints the answer and assigns the value to a variable called **ans**. If you want to perform further calculations with the answer, you can use the variable **ans** rather than retype the answer. For example, you can take the square root of the previous answer as follows:

```
>> sqrt(ans)
ans =
   7.9057
```

Notice that MATLAB assigns a new value to **ans** with each calculation. To do more complex calculations, you can assign computed values to variables of your choosing. For example:

```
>> a = cos(1)
a =
   0.5403
```

```
>> b = sin(1)
b =
   0.8415
>> a^2 + b^2
ans =
   1
```

MATLAB uses double precision floating point arithmetic, which is accurate to approximately 15 digits; however, MATLAB displays only 5 digits by default. To display more digits, type **format long**. Then all subsequent numerical output will have 15 digits displayed. Type **format short** to return to 5 digit display.

3.2 Recovering from Problems

Inevitably, when using any mathematical software system, you are bound to encounter minor glitches. Even while entering simple arithmetic commands, you may accidentally mistype an entry or inadvertently violate a MATLAB rule. In this brief section, we discuss two methods for coping with these kinds of problems.

3.2.1 Errors in Input

If you make an error in an input line, MATLAB will print an error message. For example, here's what happens when you try to evaluate **2a**:

```
>> 2a
???  2a
      |
Error:  Missing MATLAB operator.
```

The error is a missing multiplication operator *****. The correct input would be **2*a**. Note that MATLAB places a marker (a vertical line segment) at the place where it thinks the error might be; however, the actual error may have occurred earlier or later in the expression. Missing multiplication operators, brackets, and parentheses are the most common errors.

You can edit an input line by using the up-arrow key to redisplay the previous command, editing the command using the left-arrow key, and then pressing ENTER or RETURN. The up-arrow key allows you to scroll back through all the commands you've typed in a MATLAB session, and is very useful when you want to correct, modify, and reenter a previous command.

3.2.2 Aborting Calculations

If MATLAB gets hung up in a calculation, or seems to be taking too long to perform an operation, you can usually abort it by typing CTRL+C. While not foolproof, holding down the key labelled CTRL, or CONTROL, and pressing C is the method of choice when MATLAB is not responding.

3.3 Symbolic Computation

Using MATLAB's Symbolic Toolbox, you can carry out symbolic calculations such as factoring polynomials or solving algebraic equations. Type **help symbolic** to make sure that the Symbolic Toolbox is installed on your system.

To perform symbolic computations, you must use **syms** to declare the variables you plan to use to be symbolic variables. Consider the following series of commands:

```
>> syms x y
>> (x + y)*(x + y)^2
ans =
(x+y)^3
>> expand(ans)
ans =
x^3+3*x^2*y+3*x*y^2+y^3
>> factor(ans)
ans =
(x+y)^3
```

Note that symbolic output is left-justified, while numeric output is indented. This feature is often useful in distinguishing symbolic output from numerical output.

MATLAB often makes minor simplifications to the expressions you type but does not make any big changes unless you tell it to. You can use MATLAB's **simplify** command to try to express an expression as simply as possible. For example:

```
>> simplify((x^2 - y^2)/(x - y))
ans =
x+y
>> syms z
>> simplify(exp(x + log(y*exp(z))))
ans =
y*exp(x+z)
```

3.3.1 Substituting in Symbolic Expressions

When you work with symbolic expressions you often need to substitute a numerical value, or even another symbolic expression, for one (or more) of the original variables in the expression. This is done by using the **subs** command. For example, presuming the symbol expression **w** is defined and it involves the symbolic variable **u**, then the command **subs(w, u, 2)** will substitute the value 2 for the variable **u** in the expression **w**. More examples:

```
>> c = 1, syms u v
c =
    1
>> w = u^2 - v^2
w =
u^2-v^2
```

```
>> subs(w, u, 2)
ans =
4-v^2
>> subs(w, v, c)
ans =
u^2-1
>> subs(w, v, u + v)
ans =
u^2-(u+v)^2
>> simplify(ans)
ans =
-2*u*v-v^2
```

Note that in the first input line we separated two commands by a comma. Only one of them generated output, but had both resulted in an evaluation, we would see two output lines. You can enter several commands on a single line if you separate them by commas.

3.3.2 Symbolic Expressions and Variable Precision Arithmetic

As we have noted, MATLAB uses floating point arithmetic for its calculations. Using the Symbolic Toolbox, you can also do exact arithmetic with symbolic expressions. Consider the following example:

```
>> sin(pi)
ans =
   1.2246e-16
```

The answer is written in floating point format ("scientific notation") and means 1.2246×10^{-16}. However, we know that $\sin(\pi)$ is really equal to 0. The inaccuracy is due to the fact that typing **pi** in MATLAB gives an approximation to π accurate to about 15 digits, not its exact value. To compute an exact answer, instead of an approximate answer, we must create an exact *symbolic* representation of π by typing **sym('pi')**. So, to take the sine of the symbolic representation of π:

```
>> sin(sym('pi'))
ans =
0
```

This is the expected answer.

The quotes around **pi** in **sym('pi')** create a *string* consisting of the characters **pi** and prevent MATLAB from evaluating **pi** as a floating point number. The **sym** command converts the string to a symbolic expression.

The commands **sym** and **syms** are closely related. In fact, **syms x** is equivalent to **x = sym('x')**. While **syms** has a lasting effect on its argument (it declares it to be symbolic from now on), **sym** has only a temporary effect unless you assign the output to a variable, as in **x = sym('x')**.

You can also do variable precision arithmetic with the **vpa** command. For example, to print 50 digits of $\sqrt{2}$, type:

```
>> vpa('sqrt(2)', 50)
ans =
1.4142135623730950488016887242096980785696718753769
```

If you don't specify the number of digits, the default setting is 32. You can change the default with the command **digits**.

3.4 Vectors

A vector is an ordered list of numbers. You can enter a vector of any length in MATLAB by typing a list of numbers, separated by commas or spaces, inside square brackets. For example:

```
>> X = [1, 2, 3]
X =
   1     2     3
>> Y = [4 6 5 2 8]
Y =
   4     6     5     2     8
```

Suppose you want to create a vector of values running from 1 to 9. Here's how to do it without typing each number:

```
>> X = 1:9
X =
   1     2     3     4     5     6     7     8     9
```

The notation **1:9** is used to represent a vector of numbers running from 1 to 9 in increments of 1. The increment can be specified as the middle of three arguments.

```
>> X = 1:2:9
X =
   1     3     5     7     9
```

You can also use fractional or negative increments, as in **0:0.1:1** or **100:-1:0**. Finally, to extract a component of a vector, say the fourth component in **X**, type **X(4)**.

To change **X** from a row vector to a column vector, put a prime (**'**) after **X**.

```
>> X'
ans =
   1
   3
   5
   7
   9
```

You can perform mathematical operations on vectors. For example, to square the elements of the vector **X**, type:

```
>> X.^2
ans =
   1     9    25    49    81
```

The period in this expression is very important; it says that the numbers in **X** should be squared individually, or *element-by-element*. Typing **X^2** would tell MATLAB to use matrix multiplication to multiply **X** by itself and would produce an error message in this case. (We will discuss matrices in Chapter 8.) Similarly, you must type **.*** or **./** if you want to multiply or divide vectors element-by-element. For example, to multiply the elements of the vector **X** by the corresponding elements of the vector **Y**, type:

```
>> X.*Y
ans =
   4    18    25    14    72
```

Most MATLAB operations are automatically performed element-by-element. For example, you do not type a period for addition and subtraction, and you can type **exp(X)** to get the exponential of each number in **X** (the matrix exponential function is **expm**). One of the strengths of MATLAB is its ability to efficiently perform operations on vectors.

3.4.1 Suppressing Output

Typing a semicolon at the end of a MATLAB command suppresses printing of the output. The semicolon should generally be used when defining large vectors or matrices (such as **X = -1:0.1:2;**). It can also be used in any other situation where the MATLAB output need not be displayed. You can also use a semicolon, like a comma, to separate MATLAB commands on the same input line. The only difference between them is that the semicolon suppresses output and the comma does not.

3.5 Functions

MATLAB allows you to use both built-in functions and functions you define yourself.

3.5.1 Built-in Functions

MATLAB has all of the usual "elementary functions" built in. For example, **exp(x)** (*not* **e^x**) is the exponential function of **x**, and **log(x)** (*not* **ln(x)**) is the natural logarithm of **x**:

```
>> log(10)
ans =
   2.3206
```

Other built-in functions include **sqrt**, **cos**, **sin**, **tan**, and **atan** (for arctan).

3.5.2 User-defined Functions

In this section we will examine two methods to define your own functions in MATLAB: using the command **inline**, and using the operator **@** to create what is called an "anonymous function." The latter method is new in MATLAB 7 (R14), and is the preferred method for users of MATLAB 7. We shall mention **inline** periodically for the sake of users of earlier versions, but we strongly recommend that users of MATLAB 7, and users of older versions when they upgrade to R14, adopt **@** as the usual method for defining their functions. Functions can also be defined in separate files called M-files—see Chapter 4.

Here's how to define the function $f(x) = x^2$ using these commands.

```
>> f = @(x) x^2
f =
   @(x) x^2
```

Alternatively,

```
>> f1 = inline('x^2', 'x')
f1 =
   Inline function:
   f1(x) = x^2
```

Once the function is defined—by either method—you can evaluate it:

```
>> f(4)
ans =
   16
>> f1(4)
ans =
   16
```

As we observed earlier, most MATLAB functions can operate on vectors as well as scalars. To insure that your user-defined function can act on vectors, insert dots before the mathematical operators *, / and ^. Thus to obtain a vectorized version of $f(x) = x^2$, either type

```
>> f = @(x) x.^2
```

or else

```
>> f1 = inline('x.^2', 'x')
```

Now we can evaluate either function on a vector, for example

```
>> f(1:5)
ans =
   1     4     9    16    25
```

You can also use these methods to define a function of several variables. For example, either of the following

```
>> g = @(x, y) x^2 + y^2; g(1, 2)
>> g1 = inline('x^2 + y^2', 'x', 'y'); g1(1, 2)
```

results in the answer 5. If instead you define

```
>> g = @(x, y) x.^2 + y.^2;
```

then you can evaluate on vectors; thus

```
>> g([1 2], [3 4])
ans =
    10    20
```

gives the values of the function at the points (1,3) and (2,4).

3.6 Managing Variables

We have now encountered three different classes of MATLAB data: floating point numbers, strings, and symbolic expressions. In a long MATLAB session it may be hard to remember the names and classes of all the variables you have defined. You can type **whos** to see a summary of the names and types of your currently defined variables. Before you do that, please type the following input:

```
>> x = pi; y = 'pi'; z = sym('pi');
```

This makes **x** a numerical approximation to π, **y** a string expression consisting of the two letters *p* and *i*, and **z** a symbolic expression, π. Now type **whos**. The output for the MATLAB session presented in this chapter should then look as follows:

```
>> whos
  Name       Size          Bytes  Class

  X          1x5              40  double array
  Y          1x5              40  double array
  a          1x1               8  double array
  ans        1x2              16  double array
  b          1x1               8  double array
  c          1x1               8  double array
  f          1x1              16  function_handle array
  f1         1x1             824  inline object
  g          1x1              16  function_handle array
  g1         1x1             882  inline object
  u          1x1             126  sym object
  v          1x1             126  sym object
  w          1x1             138  sym object
  x          1x1               8  double array
  y          1x2               4  char array
  z          1x1             128  sym object

Grand total is 138 elements using 2388 bytes
```

We see that there are currently sixteen assigned variables in our MATLAB session representing five different data classes. Four are of class "sym object"; *i.e.*, they are symbolic objects. The variables **u** and **v** are symbolic because we declared them to be so using **syms**, **w** was defined in terms of **u** and **v**, and **z** was defined using **sym**. Seven are of class "double array"; *i.e.*, they are arrays of double precision numbers. In this case most of the arrays are of size 1×1, *i.e.*, scalars. The variables **X**, **Y**, **a**, **b**, **c** and **x** are of this size, but **ans**, the result of the last command in the previous section, is a 1×2 array. The variable **y** is of class "char array"; *i.e.*, a string expression. The last two data classes are represented by **f1** and **g1**, which are of class "inline object," and by **f** and **g**, which are of class "function handle array". The "Bytes" column shows how much computer memory is allocated to each variable.

While **whos** shows information about all defined variables, it does not show the values of the variables. To see the value of a variable, simply type the name of the variable and press ENTER or RETURN.

MATLAB commands expect particular classes of data as input, and it is important to know what class of data is expected by a given command; the help text for a command usually indicates the class or classes of input it expects. The wrong class of input usually produces an error message or unexpected output. For example, see what happens if you type **sin('pi');** the error message you see occurs because you supplied a string to a function that isn't designed to accept strings. Incidentally, string output, like symbolic output and unlike numeric output, is not indented.

To clear all defined variables, type **clear** or **clear all**. You can also type, for example, **clear x y** to clear only **x** and **y**. It is safest to clear variables before starting a new calculation. Otherwise values from a previous calculation can creep into the new calculation by accident.

3.7 Solving Equations

Before you solve differential equations with MATLAB, it is helpful to learn how to solve algebraic equations. To solve the quadratic equation $x^2 + 2x - 4 = 0$, type:

```
>> solve('x^2 + 2*x - 4 = 0')
ans =
  5^(1/2)-1
 -1-5^(1/2)
```

Here the equation to be solved is specified as a string; *i.e.*, it is surrounded by single quotes. The input to **solve** can also be a symbolic expression, but then MATLAB requires that the right-hand side of the equation be 0, and in fact the syntax for solving $x^2 - 3x = -7$ is:

```
>> syms x; solve(x^2 - 3*x + 7)
ans =
 3/2+1/2*i*19^(1/2)
 3/2-1/2*i*19^(1/2)
```

The answer consists of the exact (symbolic) solutions $(3 \pm \sqrt{19}i)/2$ (complex numbers, where the letter **i** in the answer stands for the imaginary unit $\sqrt{-1}$). To get numerical solutions, type **double(ans)**, or **vpa(ans)** to display more digits.

The **solve** command can solve higher-degree polynomial equations, as well as many other types of equations. It can also solve equations involving more than one variable. If there are fewer equations than variables, you should specify (as strings) which variable(s) to solve for. For example, type **solve('x + log(y) = 3', 'y')** to solve $x + \log y = 3$ for y in terms of x. You can specify more than one equation. For example:

```
>> [x, y] = solve('x + y^2 = 2', 'y - 3*x = 7')
x =
 -43/18+1/18*157^(1/2)
 -43/18-1/18*157^(1/2)
y =
 -1/6+1/6*157^(1/2)
 -1/6-1/6*157^(1/2)
```

This system of equations has two solutions. MATLAB reports its results by giving the two x values and the two y values for those solutions. Thus the first solution consists of the first value of x together with the first value of y. You can extract these values by typing **x(1)** and **y(1)**.

```
>> x(1)
ans =
-43/18+1/18*157^(1/2)
>> y(1)
ans =
-1/6+1/6*157^(1/2)
```

The second solution can be extracted with **x(2)** and **y(2)**.

Note that in the preceding use of the **solve** command, we assigned the output to the vector **[x, y]**. If you use **solve** on a system of equations without assigning the output to a vector, then MATLAB does not automatically display the values of the solution.

```
>> sol = solve('x + y^2 = 2', 'y - 3*x = 7')
sol =
x: [2x1 sym]
y: [2x1 sym]
```

To see the vectors of x and y values of the solution, type **sol.x** and **sol.y**. To see the individual values, type **sol.x(1)**, **sol.y(1)**, *etc.*

Some equations cannot be solved symbolically, and in these cases **solve** tries to find a numerical answer. For example:

```
>> solve('cos(x) = x')
ans =
.73908513321516064165531208767387
```

Sometimes there is more than one solution, and you may not get what you expected. For example:

```
>> solve('exp(-x) = sin(x)')
ans =
-2.0127756629315111633360706990971+2.7030745115909622139316148044265*i
```

The answer is a complex number. Though it is a valid solution of the equation, there are also real number solutions. The graphs of $\exp(-x)$ and $\sin(x)$ are shown in Figure 3.1; each intersection of the two curves represents a solution of the equation $e^{-x} = \sin(x)$.

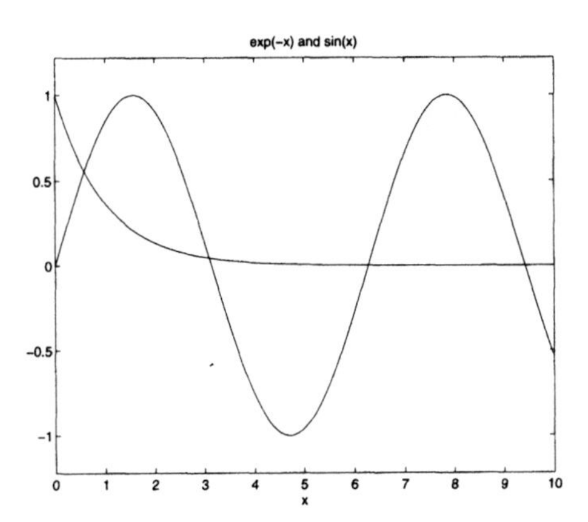

Figure 3.1: Two Intersecting Curves

You can numerically find the (approximate) solutions shown on the graph with the **fzero** command, which looks for a zero of a given function near a specified value of x. A solution of the equation $e^{-x} = \sin(x)$ is a zero of the function $e^{-x} - \sin(x)$, so to find an approximate solution near $x = 0.5$, type:

```
>> h = @(x) exp(-x) - sin(x);
>> fzero(h, 0.5)
ans =
    0.5885
```

Replace **0.5** with **3** to find the next solution, and so forth.

3.8 Graphics

In this section, we introduce MATLAB's basic plotting commands and show how to use them to produce a variety of plots.

3.8.1 Graphing with ezplot

The simplest way to graph a function of one variable is with **ezplot**, which expects a string, a symbolic expression, or a function representing the function to be plotted. For example, to graph x^2 on the interval -1 to 2 (using the string form of **ezplot**), type:

```
>> ezplot('x^2', [-1 2])
```

The graph will appear on the screen in a new window labeled "Figure 1".

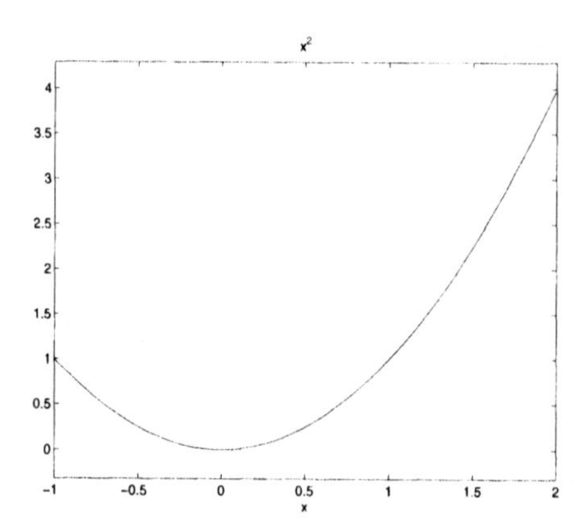

Figure 3.2: The Parabola $y = x^2$ on the Interval [-1, 2]

We mentioned that **ezplot** accepts either a string argument or a symbolic expression. Using a symbolic expression, you can produce the graph in Figure 3.2 with the following input:

```
>> syms x; ezplot(x^2, [-1 2])
```

And finally, you can use an anonymous function as the argument to **ezplot**, as in

```
>> ezplot(@(x) x.^2, [-1 2])
```

3.8.2 Modifying Graphs

You can modify a graph in a number of ways. You can change the title above the graph in Figure 3.2 by typing (in the command window, not the figure window):

```
>> title 'A Parabola'
```

The same change can be made directly in the figure window by selecting **Axes Properties...** from the **Edit** menu at the top of the figure window. (Just type the new title in the box marked "Title.") You can add a label on the vertical axis with **ylabel** or change the label on the horizontal axis with **xlabel**. Also, you can change the horizontal and vertical ranges of the graph with the **axis** command. For example, to confine the vertical range to the interval from 0 to 3, type:

```
>> axis([-1 2 0 3])
```

The first two numbers are the range of the horizontal axis; both ranges must be included, even if only one is changed.

To make the shape of the graph square, type **axis square**; this also makes the scale the same on both axes if the x and y ranges have equal length. For ranges of any lengths, you can force the same scale on both axes without changing the shape by typing **axis equal**. Generally, this command will expand one of the ranges as needed. However, if you unintentionally cut off part of a graph, the missing part is not forgotten by MATLAB. You can readjust the ranges with an **axis** command like the one above, or type **axis**

tight to automatically set the ranges to include the entire graph. Type **help axis** for more possibilities. (Remember to type **more on** first if you want to read a screenful at a time.)

You can make some of these changes directly in the figure window as you will see if you explore the pull-down menus on the figure window's tool bar. Our experience is that doing so via MATLAB commands in the command window provides more robustness, especially if you want to save your commands in an M-file (see Chapter 4) in order to reproduce the same graph later on. To close the figure, type **close** or **close all**, or simply click on the $\times$ in the upper right corner of the window.

See Section 8.5 for more ways to manipulate graphs.

3.8.3 Graphing with plot

The **plot** command works on vectors of numerical data. The syntax is **plot(X, Y)** where **X** and **Y** are vectors of the same length. For example:

```
>> X = [1 2 3]; Y = [4 6 5]; plot(X, Y)
```

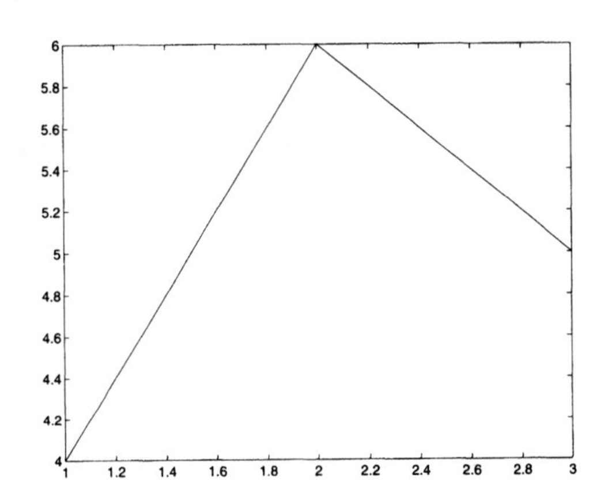

Figure 3.3: Plotting Line Segments

The **plot** command considers the vectors **X** and **Y** to be lists of the x and y coordinates of successive points on a graph, and joins the points with line segments. So, in Figure 3.3, MATLAB connects $(1,4)$ to $(2,6)$ to $(3,5)$.

To plot x^2 on the the interval from -1 to 2 we first make a list **X** of x values, and then type **plot(X, X.^2)**. We need to use enough x values to ensure that the resulting graph drawn by "connecting the dots" looks smooth. We'll use an increment of 0.1. Thus a recipe for graphing the parabola is:

```
>> X = -1:0.01:2;
>> plot(X, X.^2)
```

The result appears in Figure 3.4. Note that we used a semicolon to suppress printing of the 31-element vector **X**.

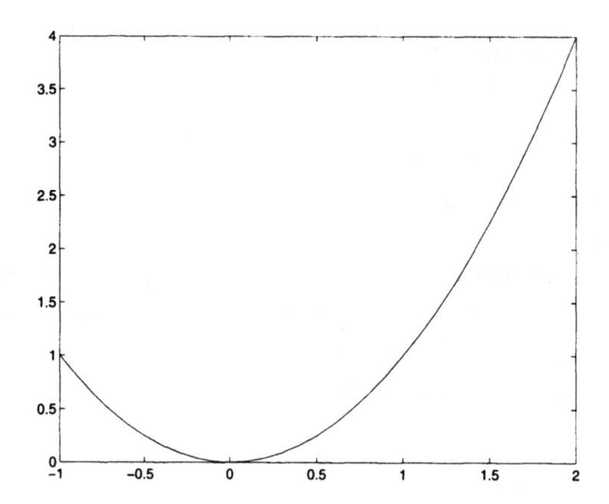

Figure 3.4: Plot of a Parabola

3.8.4 Plotting Multiple Curves

Each time you execute a plotting command, MATLAB erases the old plot and draws a new one. If you want to overlay two or more plots, use the command **hold on**. This command instructs MATLAB to retain the old graphics and draw any new graphics on top of the old. Here's an example using **ezplot**:

```
>> ezplot('exp(-x)', [0 10])
>> hold on
>> ezplot('sin(x)', [0 10])
>> hold off
>> title 'exp(-x) and sin(x)'
```

The result is shown in Figure 3.1 earlier in this chapter. The commands **hold on** and **hold off** work with all graphics commands.

Using the **plot** command, you can plot multiple curves directly. For example:

```
>> X = 0:0.01:10;
>> plot(X, exp(-X), X, sin(X))
```

Note that the vector of x coordinates must be specified once for each function being plotted.

3.8.5 Parametric Plots

Now consider another example that illustrates the flexibility of **plot**. The circle of radius 1 centered at $(0, 0)$ can be expressed in *parametric* form as $x = \cos(2\pi t), y = \sin(2\pi t)$, where t runs from 0 to 1. Though y is not expressed as a function of x, you can easily graph this curve with **plot** as follows:

```
>> T = 0:0.1:1;
>> plot(cos(2*pi*T), sin(2*pi*T))
>> axis square
```

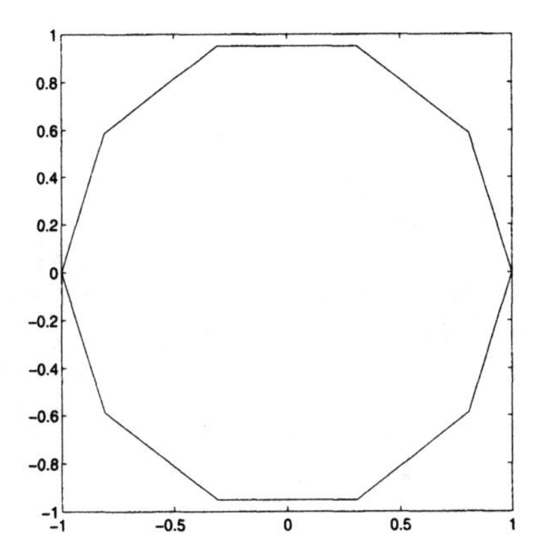

Figure 3.5: Poor Rendition of Unit Circle Due to Insufficient **plot** Points

In the plot produced by these commands (Figure 3.5), the corners between successive line segments are easily visible, despite the small increment of 0.1. In such a situation you should repeat the process with a smaller increment until you get a graph that looks smooth. An increment of 0.01 suffices here:

```
>> T = 0:0.01:1;
>> plot(cos(2*pi*T), sin(2*pi*T))
>> axis square
```

The result is shown in Figure 3.6.

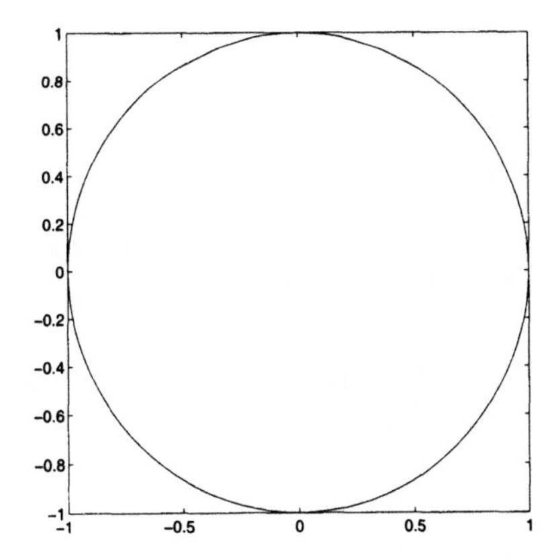

Figure 3.6: The Unit Circle

Parametric plots will be used in Chapters 13 and 14 to produce phase portraits of systems of differential equations.

3.8.6 Contour Plots

A contour plot of an expression in two variables is a plot of the *level curves* of the expression, *i.e.*, sets of points in the x-y plane where the expression has constant value. For example, the level curves of the expression x^2+y^2 are circles, and the *levels* are the squares of the radii of the circles. Contour plots are produced in MATLAB with **meshgrid** and **contour**. First use **meshgrid** to produce a grid of points in a specified rectangular region, with a specified spacing. This grid is used by **contour** to produce a contour plot in the specified region.

We can make a contour plot of $x^2 + y^2$ as follows:

```
>> [X Y] = meshgrid(-3:0.01:3, -3:0.01:3);
>> contour(X, Y, X.^2 + Y.^2)
>> axis square
```

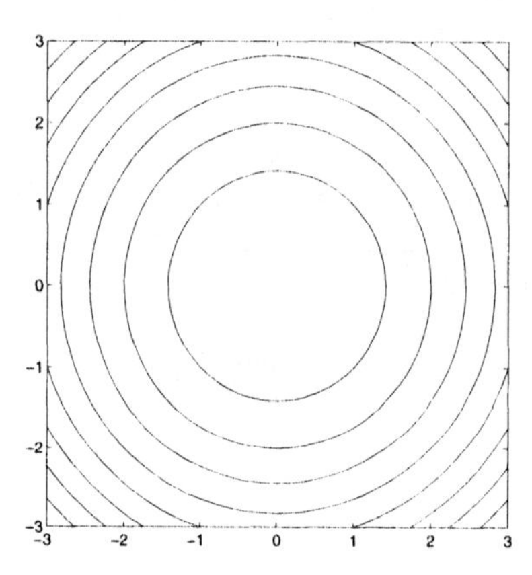

Figure 3.7: Contour Plot of $x^2 + y^2$

The plot is shown in Figure 3.7. We have used **meshgrid** and MATLAB's colon notation to produce a grid with spacing 0.01 in both directions. We have also used **axis square** to force the same scale on both axes.

You can specify *how many* level sets to plot by adding an optional integer argument to **contour**, or *what particular* level sets to plot by including an additional vector argument. For example, to plot the circles of radius 1, $\sqrt{2}$ and $\sqrt{3}$, type

```
>> contour(X, Y, X.^2 + Y.^2, [1 2 3])
```

To distinguish a vector argument from a scalar argument, the vector argument must contain at least two elements. Thus if you want to plot a single level set, you should specify the same level twice. For example, to plot the circle of radius 1, type

```
>> contour(X, Y, X.^2 + Y.^2, [1 1])
```

Replacing **[1 1]** by **1** here would give you one contour, though not this particular one. Contour plots will be used to plot implicit solutions of differential equations.

3.9 Calculus

The **diff** command differentiates a symbolic expression, and the **int** command performs integration. For example, to differentiate x^n, you could type:

```
>> syms x n
>> diff(x^n, x)
ans =
x^n*n/x
```

The answer is correct, but is not written in the form you may be used to, nx^{n-1}. You can simplify the answer as follows:

```
>> simplify(ans)
ans =
x^(n-1)*n
```

To integrate, use the **int** command:

```
>> int(cos(x), x)
ans =
sin(x)
```

(If **x** had not been declared to be a symbolic variable earlier, you would have to type **syms x** before the command above.) This is an example of an indefinite integral. You can also compute a definite integral as follows:

```
>> int(x^2*exp(-x), x, -5, 5)
ans =
-37*exp(-5)+17*exp(5)
```

Some integrals cannot be done symbolically, and in these cases MATLAB returns the unevaluated integral:

```
>> hardintegral = int(log(1+x^2)*exp(-x^2), x, 0, 1)
Warning:  Explicit integral could not be found.
> In sym.int at 58

hardintegral =
int(log(1+x^2)*exp(-x^2)),x = 0 .. 1)
```

In that case you can use MATLAB to compute a numerical approximation to the definite integral in one of two ways. First, MATLAB has two built-in numerical integration routines, **quad** and **quadl**. The one we will use here is **quadl**. Thus,

```
>> quadl(@(x) log(1+x.^2).*exp(-x.^2), 0, 1)
ans =
    0.1539
```

Here is the second method using **double** or **vpa**:

```
>> double(hardintegral)
ans =
    0.1539
```

In the second method MATLAB uses the *Symbolic Math Toolbox* to numerically evaluate the integral.

MATLAB can also deal with improper integrals. For instance, you can integrate to ∞, which is written **Inf** in MATLAB:

```
>> int(sin(x)/x, x, 0, Inf)
ans =
1/2*pi
```

Even integrals with singularities can often be evaluated:

```
>> improperintegral = int(log(atan(x)), x, 0, 1)
Warning:  Explicit integral could not be found.
> In sym.int at 58

improperintegral =
int(log(atan(x)),x = 0 .. 1)
```

But

```
>>  quadl(@(x) log(atan(x)), 0, 1)
ans =
   -1.0905
```

The command **double(improperintegral)** yields the same answer.

3.10 Some Tips and Reminders

We have introduced quite a few commands and concepts in this chapter. It will take you a while to become comfortable and proficient with them. In particular, the use of strings and symbolic expressions is often quite subtle and can trip up even experienced users. The best way to learn is through experience and by using online help (see Section 2.5) to access MATLAB documentation when necessary.

Keep in mind that previously defined (and forgotten) variables are often the source of mysterious errors. Use **whos** to keep track of defined variables. Use **clear** or **clear all** to clear old variables and functions, and use **close** or **close all** to close old figure windows. Finally, remember that typing CTRL+C will usually abort a MATLAB command. Sometimes, CTRL+C will leave MATLAB in a confused state that will interfere with future calculations. This can often be fixed by typing **clear all** to clear all variable and function definitions.

Chapter 4

Using M-files and M-books

In this chapter we describe effective procedures for working with MATLAB, and for preparing and presenting the results of a MATLAB session.

4.1 The MATLAB Desktop

Effective use of MATLAB requires understanding the various components of the MATLAB Desktop. While you can customize your Desktop to suit your own needs, we will assume in this chapter that you have retained the default configuration, with the Workspace Browser and Current Directory Browser in the upper left, the Command History in the lower left, and the Command Window on the right.

4.1.1 The Workspace

The Workspace Browser displays a table that contains, for each currently defined variable, its name, its size, and its class. The values of the variables are not displayed by default, but if you double-click on a variable in the Workspace Browser, an Array Editor will pop open. In all cases this will display the value of the variable. For numerical arrays of class "double" (and some other classes of variables), you can also change the entries of the array directly in the editor, without using the Command Window, using the menu items in the Array Editor toolbar. You can remove a variable from the Workspace by selecting it in the Workspace Browser and then pressing the "Delete" button.

If you must interrupt a session, you can avoid the need to recompute everything later by saving the current Workspace. Simply type **save myfile** to save the values of all currently defined variables in a file called `myfile.mat`. To save only the values of the variables **X** and **Y**, type

```
>> save myfile X Y
```

When you start a new session and want to recover the values of those variables, use the **load** command. For example, typing **load myfile** restores the values of all the variables stored in the file `myfile.mat`.

4.1.2 The Current Directory and MATLAB Path

We recommend that you establish a folder or directory where you can store all your MATLAB files. For example, on a *Windows* machine, you could create a directory called `C:\MatlabFiles`. Another possibility is to use the `work` folder created when MATLAB is installed, which is the default startup directory on *Windows* machines. (However, on a shared computer, you probably need to have your own individual folder.) Some people prefer to create several different directories (with more informative names) to store separate projects.

You now have two choices. When you are working on a particular project, you can tell MATLAB to use that project directory as its current working directory. This is the directory into which any files you create during your session are saved (unless you specify otherwise). To set the current directory from the command line, use the **`cd`** command. Thus on a *Windows* machine, you could type

```
>> cd C:\MatlabFiles
```

Alternatively, you can add a directory to the path list by using **`addpath`**:

```
>> addpath C:\MatlabFiles
```

MATLAB can only find the files in a directory if you either make it the current working directory or add it to the path list. When you add a directory to the path, the files it contains remain available for all your projects. The potential disadvantage of this is that you must be careful when naming files. When MATLAB searches for files, it uses the first file with the correct name that it finds in the path list, starting with the current working directory. If you use the same name for different files in different projects, you can run into problems. (You can change directories or add folders to the MATLAB path automatically every time you start MATLAB by putting the commands into a file called `startup.m` in the directory from which you launch MATLAB; see the discussion of script M-files later in Section 4.2.1.)

The easiest way to control the current directory and the path is by using the Current Directory Browser and the Set Path dialog box. The Current Directory Browser is part of the Desktop; in the default configuration, it is one of the two windows in the upper left, identified with a tab. The **Set Path** dialog box is opened from the **File** menu at the top left of the Desktop, by going to the menu item **Set Path...**. Operation of these two browsers is largely self-explanatory. The default MATLAB path includes all the `toolbox` folders for all the toolboxes installed on your machine. If there are many such toolboxes, searching for files can be slow, especially when using **`lookfor`**, and so you can speed up operation by removing from the path those toolboxes that you don't need at the moment. If you need them later, you can always put them back into the path. Note that when you add a folder to the search path, you can either just add it by itself or else use the **Add with Subfolders...** button.

The information displayed in the Set Path dialog box can also be obtained from the command line. To see the current working directory, type **`pwd`**. To list the files in the current directory type either **`ls`** or **`dir`**. To see the current path list that MATLAB will search for files, type **`path`**.

4.1.3 The Command History

Suppose you want to solve Problem 2 in Problem Set A of this book. In that problem, you are asked to calculate the values of

$$\sin(0.1)/0.1,\ \sin(0.01)/0.01,\ \text{and}\ \sin(0.001)/0.001$$

to 15 digits. Such a simple problem can be worked directly in the MATLAB Command Window. Here is a typical first try at a solution, together with the response that MATLAB displays in the Command Window:

```
>> x = [0.1, 0.01, 0.001];
>> y = sin(x)/x
y =
   0.9984
```

After completing a calculation, you will often realize that the result is not what you intended. In the example above, we intended to get a vector answer with three components, and we got only a single number. Furthermore, MATLAB displayed only 4 digits, not 15. To display 15 digits, we need the command **format long**, so one could simply add the line:

```
>> format long, y
```

which now gives the output

```
y =
   0.99835065813858
```

But this is still not what we want; we wanted three numbers and we only got one. As a matter of fact, you might guess that the one we got is $\sin(0.1)/0.1$, but even that is incorrect; $\sin(0.1)/0.1 = 0.99833416646828$, which differs from the above answer starting in the 5th decimal place. Instead, 0.99835065813858 is the "least squares" approximation to a "solution" of the simultaneous equations $y = \sin(x)/x$, with $x = 0.1,\ 0.01,\ 0.001$. The mistake is that we used the wrong symbol **/** for division, instead of "vectorized" division **./**. How can we efficiently repeat the calculation to get the correct answer? Of course, in this simple example we could just retype **sin(x)./x**. But retyping is time-consuming and error-prone, especially for complicated problems. And sometimes you will want to re-execute a command that you used earlier in a session, or even in a previous MATLAB session.

For these purposes, you can take advantage of the command history feature of MATLAB. Use the up and down arrow keys in the Command Window to scroll through the commands that you have used recently. When you locate the correct command line, you can use the mouse or the left and right arrow keys to move around in the command line, deleting and inserting changes as necessary, and then press the ENTER or RETURN key to tell MATLAB to evaluate the command. Alternatively, you can locate the old command you want to copy or modify in the Command History window in the lower left corner of the Desktop. (This window will include commands from *all* of your previous MATLAB sessions, at least until you reinitialize the command history list.) Then you can drag the old command into the Command Window with the mouse and modify it if you like.

4.2 M-files

For complicated problems, the simple editing tools provided by the Command Window and its history mechanism are insufficient. A much better approach is to create an M-file. MATLAB actually uses two different kinds of M-files: script M-files and function M-files. We shall illustrate the use of both types of M-files as we present different solutions to Problem 2.

M-files are ordinary text files containing MATLAB commands. You can create and modify them using any text editor or word processor that is capable of saving files as plain ASCII text, but we strongly recommend that you use MATLAB's built-in M-file editor, known as the Editor/Debugger. This editor automatically checks for certain MATLAB syntax requirements, and highlights different parts of the file in special colors (for example, comments in green, program control commands in blue, and strings in purple). You can open this editor with the "New M-file" or "Open" icons on the tool bar, the **Open...** option of the **File** menu, or by typing **`edit`** at the command prompt.

4.2.1 Script M-files

We are now going to show you how to construct a script M-file to solve Problem 2 of Problem Set A. Use the Editor/Debugger to create a file containing the following lines:

```
format long
x = [0.1, 0.01, 0.001];
y = sin(x)./x
```

Save the file with the name `problem2.m` in your working directory, or in some directory on your path, using the **Save As...** menu item in the **File** menu of the editor. You can name the file any way you like (subject to the usual naming restrictions on your operating system), but the "`.m`" suffix is mandatory. Note that the Editor/Debugger adds this suffix automatically.

You can tell MATLAB to run (or *execute*) this script by typing **`problem2`** in the MATLAB Command Window. (You must *not* type the "`.m`" extension here; MATLAB automatically adds it when searching for M-files.) The output—but not the commands that produce them—will be displayed in the Command Window. Now the sequence of commands can easily be changed by modifying the M-file `problem2.m`. For example, if you also wish to calculate $\sin(0.0001)/0.0001$, you could modify the M-file to read:

```
format long
x = [0.1, 0.01, 0.001, 0.0001];
y = sin(x)./x
```

and then run the modified script by typing **`problem2`**. Be sure to save your changes to `problem2.m` first; otherwise, MATLAB will not recognize them. Any variables that are set by the running of a script M-file will persist exactly as if you had typed them into the Command Window directly. For example, the program above will cause all future numerical output to be displayed with 15 digits. In order to revert to 4 digit format, you would have to type **`format short`**.

Echoing Commands. As mentioned above, the commands in a script M-file will not automatically be displayed in the Command Window. If you want the commands to be displayed along with the results, use the **echo** command:

```
echo on
format long
x = [0.1, 0.01, 0.001];
y = sin(x)./x
echo off
```

Adding Comments. It is important to include comments in a script M-file. These comments might explain what is being done in the calculation, or might interpret the results of the calculation. Any line in a script M-file that begins with a percent sign (**%**) is treated as a comment and is not executed by MATLAB. Here is a new version of `problem2.m` with a few comments added:

```
echo on
% Solution to Problem Set A, Problem 2

% Turn on 15 digit display
format long
x = [0.1, 0.01, 0.001];
y = sin(x)./x
% These values illustrate the fact that the limit of
% sin(x)/x as x approaches 0 is 1.
echo off
```

When adding comments to a script M-file, remember to put a percent sign at the beginning of each line. The Editor/Debugger automatically colors comments in green, to help you make sure that the percent signs are in the right place in your M-file. Comments that begin with a double percent sign (**%%**) have another function that we will explain below in Section 4.2.3. If you use **echo on** in a script M-file, then MATLAB will also echo the comments, so they will appear in the Command Window.

Structuring Script M-files. You should make sure that a script M-file is self-contained, unaffected by other variables that you might have defined elsewhere in the MATLAB session, and uncorrupted by leftover graphics. Your scripts should usually start with the **clear all** command, which ensures that previous definitions of variables do not affect the results. That way, your output depends only on the sequence of commands in the M-file. (This is not always strictly necessary; you really only need to clear those variables that you use in the M-file itself.) You should also either include the **close all** command at the beginning of an M-file that creates graphics, to close all graphics windows and start with a clean slate, or else open a new figure window with **figure** before plotting. (Otherwise, if there is an open graphics window with **hold** set to "on," then your new graphics will be added to that old figure window.)

Here is our example of a complete, careful, commented solution to Problem 2 of Problem Set A:

```
% Remove old variable definitions
clear all
% Remove old graphics windows
close all
% Display the command lines in the Command Window
echo on

% Solution to Problem Set A, Problem 2

% Turn on 15 digit display
format long

% Define the vector of values of the independent variable
x = [0.1, 0.01, 0.001];

% Compute the desired values
y = sin(x)./x
% These values illustrate the fact that the limit of
% sin(x)/x as x approaches 0 is equal to 1.
echo off
```

4.2.2 Function M-files

You often need to repeat a process several times for different input values of a parameter. For example, you can provide different inputs to a built-in function in order to find an output that meets a given criterion. As you have already seen, you can use the **@** syntax for anonymous functions (not available in MATLAB R13 or earlier), or the **inline** command to define your own functions. In many situations, however, it is more convenient, or even necessary, to define a function using an M-file instead of an anonymous or inline function. Anonymous and inline functions and M-files all allow you to extend the functionality of MATLAB by defining your own functions.

Let us return to Problem 2, where you are asked to compute the values of $\sin(x)/x$ for $x = 10^{-b}$ for several values of b. Suppose, in addition, that you want to find the smallest value of b for which $\sin(10^{-b})/(10^{-b})$ and 1 agree to 15 digits. Here is a function M-file called `sinepower.m` designed to solve that problem:

```
function y = sinepower(c)
% SINEPOWER computes sin(x)/x for x = 10^(-b),
% where b = 1, ..., c.
format long
b = 1:c;
y = (sin(10.^(-b))./10.^(-b))';
```

Like a script M-file, a function M-file is a plain text file that should reside in your MATLAB path. The first line of the file starts with the **function** command, which identifies the file as a function M-file. Note that the Editor/Debugger colors this special word in blue. The first line of the M-file specifies the name of the function and describes both its input arguments (or parameters) and its output values. In this example, the function is called **sinepower**. The file name (without the .m extension) and the function name should match. When you create this new function M-file in an untitled editor window and select **Save**, the Editor/Debugger knows to call the file sinepower.m. The function in our example takes one input argument, which is called **c** inside the M-file. It also returns one output value. When the function finishes executing, the value of **y** will be assigned to **ans** (by default) or to any other variable you choose, just as with a built-in function. It is good practice to follow the first line of a function M-file with one or more comment lines explaining what the M-file does. If you do, the **help** command will automatically retrieve this information. For example:

```
>> help sinepower
  SINEPOWER computes sin(x)/x for x = 10^(-b),
  where b = 1, ..., c.
```

The remaining lines of the M-file define the function. In this example, b is a row vector consisting of the integers from 1 to c. The vector y contains the results of computing $\sin(x)/x$ for $x = 10^{-b}$; the prime makes y a column vector. Notice that the output of the lines defining b and y is suppressed with a semicolon. In general, the output of intermediate calculations in a function M-file should be suppressed. In this example, it would be confusing to see output referring to variables **b** and **y**, which only make sense inside the function and are unrelated to any variables with similar names outside the function.

Here is an example that shows how to use this function:

```
>> sinepower(5)
ans =
   0.99833416646828
   0.99998333341667
   0.99999983333334
   0.99999999833333
   0.99999999998333
```

None of the values of b from 1 to 5 yields the desired answer, 1, to 15 digits. Judging from the output, you can expect to find the answer to the question we posed above by typing **sinepower(7)**.

4.2.3 Cells

A new feature in MATLAB 7 allows one to divide an M-file into subunits called *cells*. This is especially useful if your M-file is long or if you are going to **publish** it, as explained below in Section 4.4.3. The easiest way to divide an M-file into cells is to insert a comment line (which will serve as the "title" of the cell that follows) starting with a double percent sign (**%%**) followed by a space. If you open the M-file in the Editor/Debugger and click on **Enable Cell Mode** in the **Cell** menu, then when you click somewhere in the M-file, the cell that contains that location will be highlighted in light green. You can evaluate that cell by then selecting **Evaluate Current Cell** or pressing the "Evaluate Cell" icon. If you want to move on to the following cell, instead press the "Evaluate Cell and Advance" icon. This can be a big help if you've made a change in just one cell and do not want to run the whole script all over again. Once you have "enabled cell mode," you can also create more cells with the **Insert Cell Divider** item in the **Cell** menu.

4.3 Loops

A *loop* specifies that a command or group of commands should be repeated several times. The easiest way to create a loop is to use a **for** statement. Here is a simple example that computes and displays 10!.

```
f = 1;
for n = 2:10
    f = f*n;
end
f
```

The loop begins with the **for** statement and ends with the **end** statement. The command between those statements is executed a total of nine times, once for each value of **n** from 2 to 10. We used a semicolon to suppress intermediate output within the loop. In order to see the final output, we then needed to type **f** after the end of the loop. Without the semicolon, MATLAB would display each of the intermediate values 2!, 3!, ...

The Editor/Debugger automatically colors the commands **for** and **end** in blue. It improves readability if you indent the commands in between (as we did above); the Editor/Debugger does this automatically. If you type **for** in the Command Window, MATLAB does not give you a new prompt (**>>**) until you enter an **end** command, at which time MATLAB will evaluate the entire loop and display a new prompt.

4.4 Presenting Your Results

Presumably, you want to show others the results of your M-files. You may want to show them to your classmates or submit them to your instructor. You can present your results in several ways. You can let the reader run your M-file, or you can provide just the output. Finally, as we shall see in Sections 4.4.3 and 4.4.4, you can "publish" your M-file or incorporate your MATLAB output into a document, such as one prepared with *Word* or LaTeX. When presenting the final results, you must remember that the reader is not nearly as familiar with the M-file as you are; it is your responsibility to provide guidance. Most importantly, you should include numerous comments. Your comments should explain what is being calculated so the reader can understand your procedures and strategies. You should also interpret the results of the calculation. Obtaining a number or a graph is rarely an adequate answer to a question; you must explain what the numbers and graphs mean.

If your readers are simply going to run your M-files (all the way through in a single pass), then you should make liberal use of the **pause** command. Each time MAT-

LAB reaches a **pause** command, it stops executing the M-file until the user presses a key. Pauses should be placed after each major comment, after each graph, and after commands that produce important output. These pauses allow the viewer to read and understand your results. **Note:** An alternative to using **pause**, which is better if you are going to "publish" the M-file, is to separate the M-file into cells as explained in Section 4.2.3, and let your reader step through the file one cell at a time.

4.4.1 Presenting Graphics

As indicated in Chapter 3, graphics appear in a separate window. You can print the current figure by selecting the **Print...** item from the **File** menu in the graphics window. Alternatively, the **print** command (without any arguments) causes the figure in the current graphics window to be printed on your default printer. Since you probably don't want to print the graphics every time you run a script, you should not include a bare **print** command in an M-file. Instead, you should use a form of the **print** command that sends the output to a file. It is also helpful to give reasonable titles to your figures and to insert **pause** commands into your script so that viewers have a chance to see the figure before the rest of the script executes. For example:

```
xx = 0:0.2:2*pi;
plot(xx, sin(xx))
% Put a title on the figure.
title('Figure A:  Sine Curve')
pause
% Store the graph in the file figureA.eps.
print -deps figureA
```

The form of the **print** command used in this script does not send anything to the printer. Instead, it causes the current figure to be written in *Encapsulated PostScript* format to a file in the current working directory, called `figureA.eps`. This file can be printed later (on a PostScript printer), or it can be included in a word-processing document. If you intend to paste MATLAB graphics into a web page, it is better to save the figure as a `png` file with **print -dpng**. (You can also select **Save As...** from the **File** menu of the figure window and under "Save as type:" select "Portable Network Graphics file (*.png)".) If you intend to paste MATLAB graphics into a *Word* document (for more details about this, see Section 4.4.4), you can save the figure either as a `png` file or as a `tiff` file with **print -dpng** or **print -dtiff**. (Some *Word* installations also accept `eps` files.)

As a final example involving graphics, let's consider the problem of plotting the sine curves $\sin(x)$, $\sin(2x)$, and $\sin(3x)$ on the same set of axes. This is a typical example; we often want to plot several similar curves whose equations depend on a parameter. Here is a rudimentary script M-file solution to the problem:

```
echo on
% Define the x-values.
xx = (0:0.01:2)*pi;
% Open a new figure window for several plots.
```

```
figure, hold on
% Run a loop to plot three sine curves.
for c = 1:3
   plot(xx, sin(c*xx))
   echo off
end
echo on, hold off, axis([0, 2*pi, -1, 1])
% Put a title on the figure.
title('Several Sine Curves')
```

The result is shown in Figure 4.1.

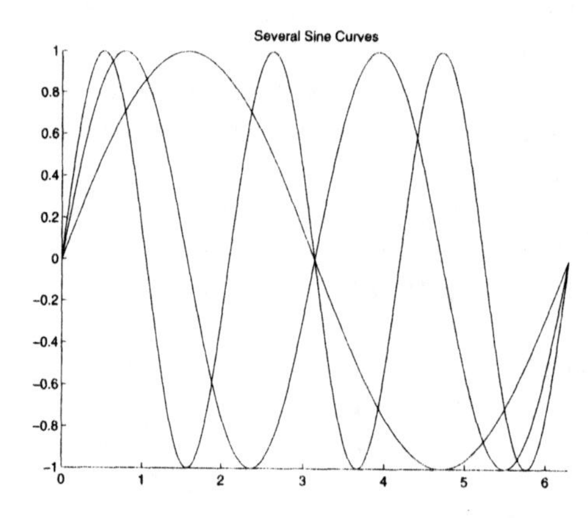

Figure 4.1: Sine Curves

Let's analyze this solution. We start by defining the x-values (as fractional multiples of π from 0 to 2π). The **`figure`** command opens a new graphics window, and the **`hold on`** command lets MATLAB know that we want to draw several curves on the same set of axes. The lines between **`for`** and **`end`** constitute a for-loop, as described above. The important part of the loop is the **`plot`** command, which plots the desired sine curves. We inserted an **`echo off`** command so that we only see the loop commands once in the Command Window. Finally, we turn echoing back on after exiting the loop, use **`hold off`** to release the graphics window, and **`title`** the figure.

4.4.2 Pretty Printing

If **`s`** is a symbolic expression, then typing **`pretty(s)`** displays **`s`** in *pretty print* format, which uses multiple lines on your screen to imitate written mathematics. The result is often more easily read than the default one-line output format. An important feature of **`pretty`** is that it *wraps* long expressions to fit within the margins (80 characters wide) of a standard-sized window. If your symbolic output is long enough to extend past the right edge of your window, it probably will be truncated when you print your output, so you should use **`pretty`** to make the entire expression visible in your printed output.

4.4.3 "Publishing" an M-file

MATLAB 7 comes with a very convenient mechanism to convert an M-file to a readable document, the **`publish`** command. Alternatively, once you have "enabled cell mode," you can also use the "publish" icon on the cell toolbar (which appears under the regular editor toolbar) in the Editor/Debugger. The default, which we have found works best, is to publish to `html` (*i.e.*, a web page), but you can also publish to a *Word* document or *PowerPoint* presentation (on a PC) or to a LaTeX document (regardless of the operating system). "Publishing" an M-file reproduces the M-file along with its command line output and whatever figure windows it generates. Each cell of the M-file appears as a separate section, under the heading taken from the appropriate line starting with a double percent sign (**`%%`**). Within the section for a cell, the MATLAB input lines are printed, followed by the output (in a lighter color). You can also format some of the text, insert bulleted lists, *etc.*, using the **Insert Text Markup** item in the **Cell** menu of the Editor/Debugger. (Remember to do this *before* publishing; otherwise it will have no effect.)

If you are going to "publish" your M-file, the **`echo`** command is unnecessary, since the text of the M-file, color-coded as in the Editor/Debugger, automatically appears. Also, the **`pause`** command is unnecessary (even a nuisance, since it interrupts the publishing until you hit ENTER or RETURN), and the **`print`** command is also unnecessary. But one additional comment about graphics is in order. If your M-file puts two successive figures in the same figure window, then only the last figure will be written to the published document, unless the figures appear in different cells. So it helps either to insert a cell divider each time a figure is created or else to precede each graphics command by the **`figure`** command, so that it appears in a new window.

4.4.4 M-books

Another sophisticated way of presenting MATLAB output, without having to produce an M-file first, is in a Microsoft *Word* document, incorporating text, MATLAB commands, and graphics. (This option is not available on *UNIX* machines, but on such computers one can paste MATLAB commands and graphics into a LaTeX document with the help of the LaTeX command **`\includegraphics`**. In fact, that is how we produced the graphics in this book!)

A simple first approach is to prepare a *Word* document with explanatory comments, and to paste in one's MATLAB commands (one can do this in another font) or include one's M-files (using **File...** from *Word*'s **Insert** menu). Finally, you can paste in the graphics using **Picture ▸ From File...** from the **Insert** menu in *Word*. You should have first saved the graphics in `png`, `tiff`, or `eps` format.

A more robust approach, which tends to be less cumbersome in the long run, is to enable M-books on your machine. An M-book is a *Word* document with embedded executable

MATLAB code (that runs as a *Word* macro, via the intermediary of *Visual Basic*). You can launch an M-book by typing **notebook** in the Command Window, or by starting *Word*, choosing **New...** from the **File** menu, and selecting `m-book.dot` as the Document template. If the file `m-book.dot` does not already exist on your computer, you need to *enable M-books* first. This is done by typing

```
>> notebook -setup
```

and following the instructions. You may be prompted for the version of *Word* that you are using. For M-books to run properly, you must enable execution of macros in *Word*. This may require that you change your *Word* "Security Level." To do this, on the **Tools** menu in *Word*, click **Options....** Click the **Security** tab. Under "Macro Security", click **Macro Security...**, and go to the "Security Level" tab.

Once you have successfully launched an M-book, it will behave just like any other *Word* document, except for the **Notebook** menu at the top. If you type a MATLAB command and hit CTRL+ENTER, or else highlight the command with the mouse and select **Evaluate Cell** in the **Notebook** menu, MATLAB will evaluate the command and send the output back to *Word*. For ease in reading, *Word* typesets "input cells" (MATLAB input) in green Courier bold and "output cells" (MATLAB output) in blue Courier type. You have an option (which you can adjust with the **Notebook Options...** item in the **Notebook** menu) of having figure windows appear separately, or having them appear in the M-book, or both.

In one respect, M-books behave like M-files; you can modify them and run them again and again. If you find you mistyped a command or want to change a command, you can simply go back to the appropriate input cell, change it, and then re-evaluate it. The new output will replace the old. No one needs to know you originally made a mistake!

4.4.5 Preparing Homework Solutions

In this section, we summarize the steps that you can use to prepare the solutions to the homework problems in this book, or for that matter, to present the results of any project you work on using MATLAB.

1. Decide whether you are going to submit an M-file (for your reader to run himself), a "published" M-file, a word-processing document with pasted in MATLAB code and output, or an M-book. Which of these you choose may depend on the configuration of your machine and the preferences of your instructor or audience.

2. Regardless of which method you choose, be sure you have explained what you are doing in comment lines (if you produce an M-file or published M-file) or accompanying text (if you produce a word-processed document or an M-book).

3. Be sure to go back and correct all errors. Try some of the debugging techniques listed below in Section 4.5. Your reader does not need to see false starts; delete them from your M-file or M-book.

If you want to create a published document from a script M-file, here is the procedure from start to finish:

1. Create an M-file in your current working directory to hold the solution. Keep the M-file open in the Editor/Debugger while you are editing it. (You might want to include **echo on** in your M-file so that you can see each command together with its output.) Select **Enable Cell Mode** and insert cell divisions as needed, especially if the file is long or creates more than one figure.

2. Continue editing and running the M-file until you are confident that it contains the MATLAB commands that solve the problem, and until it runs without errors. (See the debugging hints below.)

3. Add comments to your M-file to explain the method being used to solve the problem and to interpret the results. These comments can be formatted with the **Cell** menu. Give titles to your figures.

4. Once your M-file is ready, publish it in the document type you desire with either the **publish** command or the "publish" icon. Preview the result, and if necessary, either edit the document (for example, within *Word* or an `html` editor) or else change the M-file and publish it again.

To illustrate the results of this process, we revisit Problem 2 of Problem Set A. Here is the result of publishing (to LaTeX) a script M-file that solves the problem and adds a graphical presentation of the results. You can examine the M-file that produced this output on our web site, at

`http://www.math.umd.edu/undergraduate/schol/ode/Mfiles/`

Solution to Problem Set A, Problem 2

Contents

- Numerical Solution
- Graphical Solution

Numerical Solution

Turn on 15 digit display.

```
format long
clear all
% Assign values to the independent variable.
X = [0.1, 0.01, 0.001]
% Compute the desired values
Y = sin(X)./X
% These values illustrate the fact that the limit of
% sin(x)/x as x approaches 0 is 1.
```

```
X =

   0.10000000000000   0.01000000000000   0.00100000000000


Y =

   0.99833416646828   0.99998333341667   0.99999983333334
```

Graphical Solution

We can also illustrate the same fact graphically.

```
close all
ezplot('sin(x)/x', [0 0.1])
hold on
plot(X, Y, 'o')
title 'The Limit of sin(x)/x'
hold off
```

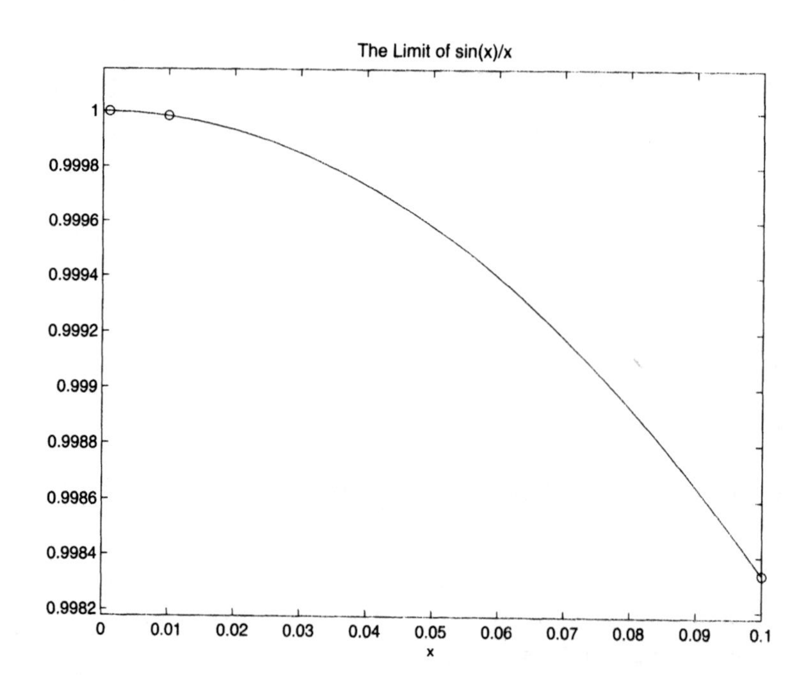

4.5 Debugging Your M-files

You can edit your M-file repeatedly until it produces the desired output. Generally, you will run the script each time you edit the file. If the program is long or involves complicated calculations or graphics, it could take a while each time. So you need a strategy for debugging. Here are some general tips:

1. The single most common problem is to forget that variable or function definitions persist until cleared, and can interfere with usages of the same variable name. If a command is giving unexpected output, try **`clear all`** and then re-execute. (This may require redefining the variables you are consciously using.)

2. Use the **`echo on`** command early in a script M-file so that you can see "cause" as well as "effect". However, use the semicolon to suppress output, especially when constructing arrays.

3. If you are producing graphics, use the **`hold`** command carefully. Use **`hold off`** as soon as you are done with a figure, so that the next figure is not inserted on top of it, and insert a **`pause`** statement or cell divider so that the next graphics command will not obliterate the current one before you have time to see it.

4. Do not include bare **`print`** commands in your M-files. Instead, print to a file.

5. The Editor/Debugger has several useful features for debugging M-files:

   (a) Splitting the M-file into cells makes it possible to debug one cell at a time.

   (b) The menu item **Check Code with M-Lint** in the **Tools** menu will check your M-file for common potential errors, such as places where you forgot a semicolon and one is usually customary. These are not always actual errors, but it is worth rechecking them. You can also get an M-lint report on an M-file by typing **`mlintrpt`** and the file name in the Command Window.

   (c) You can insert "breakpoints" into the file using the **Debug** menu or the red dot icon on the toolbar. This will stop execution of the file at these points so you can see what's going on. After you're done debugging, remember to click on the red X to clear breakpoints.

   (d) You can run your script interactively from the Editor/Debugger using **Run** from the **Debug** menu. When you get to a breakpoint, execution will stop. If you hit ENTER or RETURN, a special prompt **`K>>`** will appear in the Command Window. At this point you can type any normal MATLAB command. This is useful if you want to examine or reset some variables in the middle of a program run. You can resume execution of the M-file with the **Debug** menu, and also step through the M-file line by line with **Step** from the **Debug** menu. Alternatively, to resume the execution of your M-file, type **`return`** at the **`K>>`** prompt.

6. Finally, remember that you can stop a running M-file by typing CTRL+C. This is useful if a calculation is taking too long, if MATLAB is spewing out undesired output, or if, at a **pause** command, you realize that you want to stop execution completely.

Problem Set A

Practice with MATLAB

In this problem set, you will use MATLAB to do some basic calculations, and then to plot, differentiate, and integrate various functions. This problem set is the minimum you should do in order to reach a level of proficiency that will enable you to use MATLAB throughout the course. A solution to Problem 2 appears in Section 4.4.5, and a solution to Problem 4 appears in the *Sample Solutions*.

1. Evaluate:

   (a) $\dfrac{413}{768+295}$ (as a decimal),

   (b) 2^{123}, both as an approximate number in scientific notation and as an exact integer,

   (c) π^2 and e to 35 digits,

   (d) the fractions $\frac{61}{88}$, $\frac{13863}{20000}$, and $\frac{253}{365}$, and determine which is the best approximation to $\ln(2)$.

2. Evaluate to 15 digits:

   (a) $\dfrac{\sin(0.1)}{0.1}$,

   (b) $\dfrac{\sin(0.01)}{0.01}$,

   (c) $\dfrac{\sin(0.001)}{0.001}$.

3. Graph the equations:

   (a) $y = x^3 - x$ on the interval $-1.5 \leq x \leq 1.5$,

   (b) $y = \tan x$ on the interval $-2\pi \leq x \leq 2\pi$,

   (c) $y^2 = x^3 - x$ on the interval $-2.5 \leq x \leq 2.5$. (*Hint*: If the first argument to **`ezplot`** is a symbolic expression in two variables, MATLAB plots the locus of points where this expression is equal to 0.)

4. The **factor** command factors an expression containing symbolic variables, and also factors integers into their prime factors.

   (a) Factor $x^3 + 5x^2 - 17x - 21$.

   (b) Find the prime factorization of 987654321.

5. Plot the functions x^8 and 4^x on the same graph, and determine how many times their graphs intersect. (*Hint*: You will probably have to make several plots, using various intervals, in order to find all the intersection points.) Now find the values of the points of intersection, first using the **fzero** command and then using the **solve** command. Do these two commands produce the same answers?

6. Use the command **limit** (type **help limit** for the syntax) to compute the following limits:

   (a) $\displaystyle\lim_{x\to 0} \frac{\tan x}{x}$,

   (b) $\displaystyle\lim_{x\to 0+} \frac{1}{x}$ and $\displaystyle\lim_{x\to 0-} \frac{1}{x}$,

   (c) $\displaystyle\lim_{x\to\infty} x\,e^{-x^2}$ and $\displaystyle\lim_{x\to -\infty} x\,e^{-x}$,

   (d) $\displaystyle\lim_{x\to 0} \frac{\ln(1-x)+x}{x^2}$.

7. Compute the following derivatives using the command **diff**:

   (a) $\displaystyle\frac{d}{dx}\left(\frac{x^3}{x^2+1}\right)$,

   (b) $\displaystyle\frac{d}{dx}(\sin(\sin(\sin x)))$,

   (c) $\displaystyle\frac{d^3}{dx^3}(\arctan x)$,

   (d) $\displaystyle\frac{d}{dx}(\sqrt{1+x^2})$,

   (e) $\displaystyle\frac{d}{dx}(e^{x\ln(x)})$.

8. Compute the following integrals using the command **int**:

   (a) $\int e^{-3x}\sin x\,dx$,

   (b) $\int (x+1)\ln x\,dx$,

   (c) $\displaystyle\int_0^1 \sqrt{\frac{x}{1-x}}\,dx$,

   (d) $\displaystyle\int_{-\infty}^{\infty} e^{-x^2}\,dx$,

(e) $\int_0^1 \sqrt{1+x^4}\,dx$. For this example, also compute the numerical value of the integral.

9. Use **solve** to solve the equation

$$x^5 - 3x^2 + x + 1 = 0. \tag{A.1}$$

Find the numerical values of the five roots. Plot the graph of the 5th degree polynomial on the left-hand side of (A.1) on the interval $-2 \le x \le 2$. Now explain your results—in particular, reconcile your five roots with the fact that the graph touches the x-axis only twice. To verify this, you should include the x-axis in your graph and restrict the y-axis appropriately.

10. In one-variable calculus you learned that the local max/min of a differentiable function $y = f(t)$ are found among the *critical points* of f, that is, the points t where $f'(t) = 0$. For example, consider the polynomial function

$$y = f(t) = t^6 - 4t^4 - 2t^3 + 3t^2 + 2t$$

on the interval $[-3/2, 3/2]$.

(a) Graph $f(t)$ on that interval.

(b) How many local max/min points do you see? It's a little hard to determine what is happening for negative values of t. If need be, redraw your graph by restricting the t and/or y axis by using **axis**.

(c) Now use MATLAB to differentiate f and find the points t where $f'(t) = 0$ on the interval $[-3/2, 3/2]$. How many are there? Use **fzero** to hone in on their values.

(d) Now verify that the negative critical point is indeed an *inflection point* by graphing $f''(t)$ on the interval $-1.2 \le t \le -0.8$. How does that graph establish that the point is an inflection point?

11. (a) Use **solve** to simultaneously solve the pair of equations

$$\begin{cases} x^2 - y^2 = 1, \\ 2x + y = 2. \end{cases}$$

(b) Plot the two curves on the same graph and visually corroborate your answer from part (a). (*Hint*: To help you plot the hyperbola $x^2 - y^2 = 1$, recall that its two branches can be parameterized by the parametric equations $x = \cosh t, y = \sinh t$, and $x = -\cosh t, y = \sinh t$.)

12. MATLAB has a command **symsum** that you can use to sum infinite series.

(a) Use it to sum the famous infinite series $\sum_{n=1}^{\infty} \frac{1}{n^2}$.

(b) Use it to sum the geometric series $\sum_{n=0}^{\infty} x^n$.

(c) Define a function $f(x) = \sum_{n=1}^{\infty} \frac{x^n}{n}$. Use **symsum** to identify the function.

(d) Compute $f'(x)$ by differentiating the function you found in part (c).

(e) Now sum the series $\sum_{n=1}^{\infty} \frac{d}{dx}\left(\frac{x^n}{n}\right)$. What does this suggest about differentiating a function that is defined by a power series?

Chapter 5

Solutions of Differential Equations

In this chapter, we show how to solve differential equations with MATLAB. For many differential equations, the command **dsolve** produces the general solution to the differential equation, or the specific solution to an associated initial value problem. We also discuss the existence, uniqueness, and stability of solutions of differential equations. These are fundamental issues in the theory and application of differential equations. An understanding of them helps in interpreting and using results produced by MATLAB.

5.1 Finding Symbolic Solutions

Consider the differential equation

$$\frac{dy}{dt} = f(t, y). \tag{5.1}$$

A solution to this equation is a differentiable function $y(t)$ of the independent variable t that satisfies $y'(t) = f(t, y(t))$ for all t in some interval. For some functions f, it is possible to find a formula for the solutions to (5.1); we call such a formula a *symbolic solution* or *formula solution*. Finding a symbolic solution is generally not a straightforward task, and not surprisingly, computational algorithms for solving differential equations symbolically are imperfect. Nonetheless, MATLAB's symbolic differential equation solver **dsolve** can correctly solve most of the differential equations that can be solved with the standard solution methods one learns in an introductory course.

A symbolic solution to (5.1) can take one of several forms:

- an explicit solution that expresses y as an elementary function of t;
- an explicit solution that expresses y in terms of special functions of t;

- an implicit solution that relates y and t algebraically without expressing y as a function of t;
- a solution that relates y and t through a formula involving integrals.

By *elementary functions* we mean the standard functions of calculus: polynomials, exponentials and logarithms, trigonometric functions and their inverses, and all combinations of these functions through algebraic operations and compositions. By *special functions* we mean various non-elementary functions that mathematicians have given names to, often because they arise as solutions of particularly important differential equations.

As a practical matter, a solution from **dsolve** is most useful when it expresses y explicitly in terms of built-in MATLAB functions of t; these include the elementary functions and many special functions. In the following example, we illustrate how to use **dsolve** and its output in this case. See Section 5.4 for examples involving other types of solutions you may get with **dsolve**.

Example 5.1 Consider the linear differential equation

$$\frac{dy}{dt} = t^2 + y.$$

You can find the general solution to this equation in MATLAB by typing:

```
>> dsolve('Dy = t^2 + y', 't')
ans =
-t^2-2*t-2+exp(t)*C1
```

The solution of the differential equation is the expression following the equal sign. Notice that MATLAB produces the answer in terms of an arbitrary constant **C1**. (For higher order equations, there will be as many arbitrary constants as the order of the equation.)

You can obtain specific solutions by choosing specific values for **C1**. In particular, you can find the solution satisfying a given initial condition by imposing the initial condition on the general solution and solving for **C1**. Alternatively, you can specify the initial condition as well as the differential equation when you invoke **dsolve**. To solve the initial value problem

$$\frac{dy}{dt} = t^2 + y, \qquad y(0) = 3,$$

type:

```
>> sol1 = dsolve('Dy = t^2 + y', 'y(0) = 3', 't')
sol1 =
-t^2-2*t-2+5*exp(t)
```

In this example, we have given the name **sol1** to the output.

Next, suppose you want to plot the solution or find its value at a particular value of t. To plot the solution on the interval $0 \leq t \leq 2$, you can type **ezplot(sol1, [0 2])**. However, you can't type **sol1(2)** to get the value of the solution at $t = 2$, because the output of **dsolve** is a symbolic expression, not a function. To display the value of the solution at $t = 2$, type:

```
>> subs(sol1, 't', 2)
ans =
   26.9453
```

(Unlike the examples using **subs** in Section 3.3.1, you must put single quotes around **t** here, unless you have previously declared **t** to be a symbolic variable.) You can also substitute a vector of values for **t**; for example, **subs(sol1, 't', 0:2)** will evaluate the solution at $t = 0, 1, 2$.

If you are going to evaluate the solution many times, it may be convenient to define an anonymous function corresponding to the solution. For the current example, you can first type

```
>> y1 = @(t) -t.^2 - 2*t - 2 + 5*exp(t)
```

and then type **y1(2)** to evaluate the solution. Notice the period before the caret above, which allows **y1** to accept vector input. Alternatively, you can avoid retyping the solution by defining **y1** as follows:

```
>> y1 = @(t) eval(vectorize(sol1))
```

We discuss the commands **eval** and **vectorize** in Chapter 8.

We often want to study a family of solutions obtained by varying the initial condition. Here is a natural way to do this in MATLAB. Begin by solving the differential equation with a generic initial value. For example:

```
>> sol1a = dsolve('Dy = t^2 + y', 'y(0) = c', 't')
sol1a =
-t^2-2*t-2+exp(t)*(2+c)
```

MATLAB expresses the solution formula in terms of the initial value c. Now suppose we want to plot the solution curves with initial values $y(0) = -3, -2, \ldots, 3$ on the interval $0 \leq t \leq 2$. One approach is to type **hold on** and plot the curves one at a time. To save some typing, you can execute the same commands in a loop. The commands

```
>> figure; hold on
>> syms t
>> for cval = -3:3
       ezplot(subs(sol1a, 'c', cval), [0 2])
   end
>> axis tight
>> title 'Solutions of Dy = t^2 + y with y(0) = -3, ..., 3'
>> xlabel t, ylabel y
>> hold off
```

produce the graph in Figure 5.1. The graphics options **axis**, **title**, **xlabel**, and **ylabel** are explained in Section 3.8.2. Without **axis tight**, the range set by the last **ezplot** command would cut off part of the curves from the previous commands. Also, if we didn't supply a title, the title from the last **ezplot** command would have remained, giving (misleadingly) only the formula for one of the seven solutions.

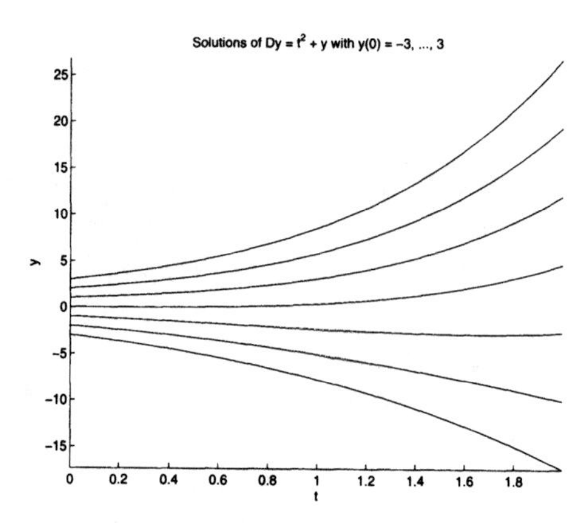

Figure 5.1: Solutions of $dy/dt = t^2 + y$

5.2 Existence and Uniqueness

The fundamental existence and uniqueness theorem for differential equations guarantees that every initial condition $y(t_0) = y_0$ leads to a unique solution near t_0, provided that the right-hand side of the differential equation (5.1) is a "nice" function. A theorem like this appears in virtually every textbook on ordinary differential equations.

Theorem 5.1 *Suppose that f and $\partial f/\partial y$ are continuous functions in the rectangle*

$$R = \{(t, y) \ : \ a \leq t \leq b, c \leq y \leq d\},$$

and that the point (t_0, y_0) lies in R. Then the initial value problem

$$dy/dt = f(t, y), \qquad y(t_0) = y_0$$

has a unique solution $y(t)$ that exists at least as long as its graph stays within R.

Notice that by virtue of being a solution, $y(t)$ is differentiable (and hence continuous) for as long as it exists, and that since f is continuous, so is dy/dt.

Graphically, this theorem says that there is a smooth solution curve (or *integral curve*) through every point in R and that the solution curves cannot cross. Thus an initial value problem (IVP) has exactly one solution, but, since there are an infinite number of possible initial conditions, a differential equation has an infinite number of solutions. This principle is implicit in the results obtained above with **dsolve**; when we do not specify an initial condition, the solution depends on an arbitrary constant; when we specify an initial condition, the solution is completely determined.

It is important to remember that the existence and uniqueness theorem only guarantees the existence of a solution *near* the initial point t_0. Consider the initial value problem

$$\frac{dy}{dt} = y^2, \qquad y(0) = 1. \tag{5.2}$$

To solve it symbolically, type:

```
>> sol = dsolve('Dy = y^2', 'y(0) = 1', 't')
sol =
-1/(t-1)
```

To understand where the solution exists, we graph **sol** (see Figure 5.2):

```
>> ezplot(sol, [-1 2])
```

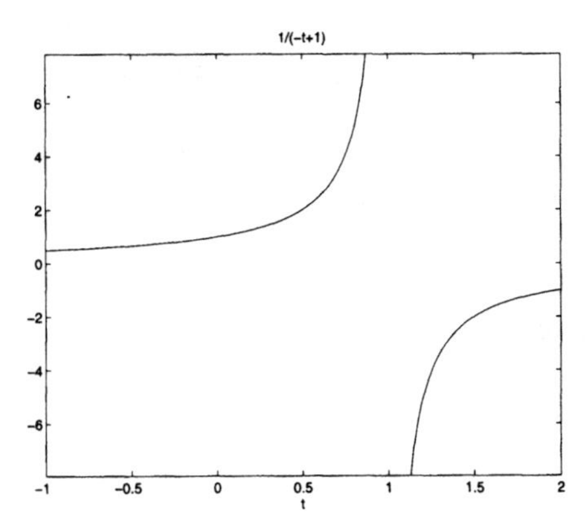

Figure 5.2: Graph of $y = 1/(1-t)$, which for $t < 1$ is the Solution of the IVP (5.2)

We see that the graph has two branches. Since the left branch passes through the initial data point $(0, 1)$, it is the desired solution. Furthermore, we see that the solution exists from $-\infty$ to 1 but does not extend beyond 1. In fact, it becomes unbounded as t approaches 1 from the left. The right branch of the graph depicts more of the function $1/(1-t)$, but it is not part of the solution of the initial value problem.

5.3 Stability of Differential Equations

In addition to existence and uniqueness, the sensitivity of the solution of an initial value problem to the initial condition is a fundamental issue in the theory and application of differential equations. When a differential equation is used to model a physical system, the exact initial condition is generally unknown; instead, there may be a small range of possible initial values. Then in order to assess the amount of uncertainty in predictions made by the model, it is important to know whether solutions that are close to each other at the initial point t_0 remain close together at other values of t. We examine this issue in the following examples.

Example 5.2 Consider the initial value problem

$$\frac{dy}{dt} + 2y = e^{-t}, \qquad y(0) = y_0.$$

We ask: How does the solution $y(t)$ depend on the initial value y_0? More specifically: Does the solution depend continuously on y_0? Do small variations in y_0 lead to small, or large, variations in the solution? Since the solution is

$$y(t) = e^{-t} + (y_0 - 1)e^{-2t},$$

we immediately see that, for any fixed t, the solution $y(t)$ depends continuously on y_0. We can say more. If we let $\tilde{y}(t)$ be the solution of the same equation, but with the initial condition $\tilde{y}(0) = \tilde{y}_0$, then $\tilde{y}(t) = e^{-t} + (\tilde{y}_0 - 1)e^{-2t}$. So, we have

$$|y(t) - \tilde{y}(t)| = |y_0 - \tilde{y}_0|e^{-2t}.$$

Thus for $t \geq 0$ we see that $|y(t) - \tilde{y}(t)|$ is never larger than $|y_0 - \tilde{y}_0|$. In fact, $|y(t) - \tilde{y}(t)|$ decreases as t increases. For $t \leq 0$ the situation is different. When t is a large negative number, the initial difference $|y_0 - \tilde{y}_0|$ is magnified by the large factor e^{-2t}. Even though $y(t)$ depends continuously on y_0, small changes in y_0 lead to large changes in $y(t)$. For example, if $|y_0 - \tilde{y}_0| = 10^{-3}$, then $|y(-7) - \tilde{y}(-7)| = 10^{-3}e^{14} \approx 1203$. These observations are confirmed by Figure 5.3, a plot of the solutions corresponding to initial values $y(0) = 0.97, 1, 1.03$ for $-3 \leq t \leq 3$.

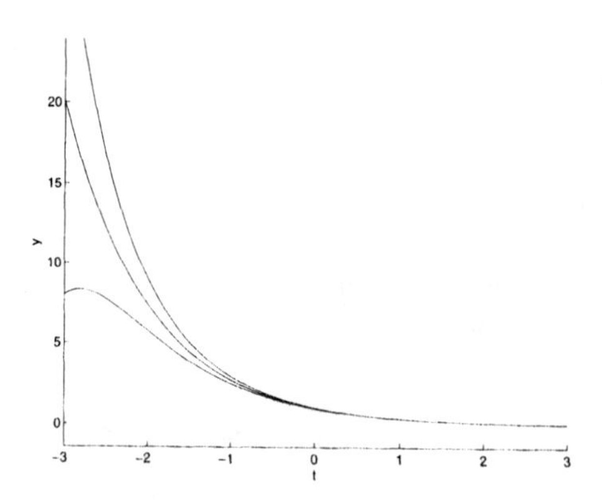

Figure 5.3: Solutions of $dy/dt + 2y = e^{-t}$

Example 5.3 The solution of the initial value problem

$$\frac{dy}{dt} - 2y = -3e^{-t}, \qquad y(0) = y_0$$

is $y(t) = e^{-t} + (y_0 - 1)e^{2t}$. Again, $y(t)$ depends continuously on y_0 for fixed t. Letting $\tilde{y}(t)$ be the solution with initial value $\tilde{y}_0$, we see that

$$|y(t) - \tilde{y}(t)| = |y_0 - \tilde{y}_0|e^{2t}.$$

Now we see that $y(t)$ is very sensitive to changes in the initial value for t large and positive, but insensitive for t negative. These observations are confirmed by Figure 5.4, a plot of the solutions corresponding to initial values $y(0) = 0.97, 1, 1.03$ for $-2 \leq t \leq 2$.

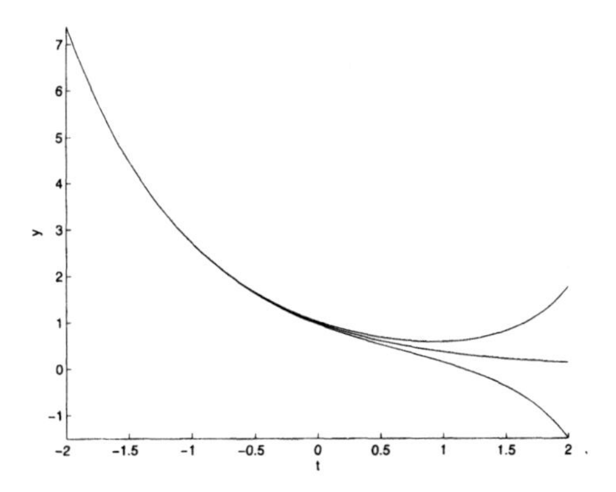

Figure 5.4: Solutions of $dy/dt - 2y = -3e^{-t}$

Often, we are primarily interested in positive values of t, such as when t corresponds to time in a physical problem. If an initial value problem is to predict the future of a physical system effectively, the solution for positive t should be fairly insensitive to the initial value—*i.e.*, small changes in the initial value should lead to small changes in the solution for positive time—for the following reason. As we mentioned at the beginning of this section, for a physical system the initial value y_0 typically is not known exactly. When it is found by measurement, the result is an approximate value $\tilde{y}_0$. Then, if $\tilde{y}(t)$ is the solution corresponding to $\tilde{y}_0$ and the solution is very sensitive to the initial value, $\tilde{y}(t)$ will have little relation to the actual state $y(t)$ of the system as t increases.

An initial value problem whose solution is fairly insensitive to small changes in the initial value as t increases is called *stable*, while an initial value problem whose solution is very sensitive to small changes in the initial value is called *unstable*. If all of the solutions of a differential equation are stable (as in Example 5.2), we say the equation is stable, and likewise if all solutions are unstable (as in Example 5.3), we say the equation is unstable. As we have seen, if an equation is unstable and is used over a long time interval, a small error in the initial value can result in a large error later on. Caution should be exercised when using an unstable equation to model a physical problem.

Remark 5.1 We have considered stability as t increases, *i.e.*, stability to the right. We can also consider stability to the left. There are equations that are stable both to the left and to the right, and equations that are unstable both to the left and right.

The following theorem, which we state without proof, is often useful in assessing stability.

Theorem 5.2 *Suppose that $f(t, y)$ has continuous first order partial derivatives in the vertical strip*

$$S = \{(t, y) : t_0 \le t \le t_1, -\infty < y < \infty\},$$

and suppose there are numbers K and L such that

$$K \le \frac{\partial f}{\partial y}(t, y) \le L, \quad \textit{for all } (t, y) \in S.$$

If $y(t)$ and $\tilde{y}(t)$ are solutions of $dy/dt = f(t, y)$ on the interval $t_0 \le t \le t_1$ with initial values $y(t_0) = y_0$ and $\tilde{y}(t_0) = \tilde{y}_0$, respectively, then

$$|y_0 - \tilde{y}_0| e^{K(t-t_0)} \le |y(t) - \tilde{y}(t)| \le |y_0 - \tilde{y}_0| e^{L(t-t_0)},$$

for all $t_0 \le t \le t_1$.

If $L \le 0$, then the right-hand inequality in the theorem shows that

$$|y(t) - \tilde{y}(t)| \le |y_0 - \tilde{y}_0|, \quad \text{for all } t_0 \le t \le t_1.$$

Thus the solutions differ by no more than the difference in the initial values, and the differential equation is stable (at least for $t_0 \le t \le t_1$). Moreover, if $L > 0$, but not too large, and $t_1 - t_0$ is not too large, then

$$|y(t) - \tilde{y}(t)| \le M |y_0 - \tilde{y}_0|, \quad \text{for all } t_0 \le t \le t_1,$$

where $M = e^{L(t_1 - t_0)}$ is a moderate-sized constant. Thus the equation is only mildly sensitive to changes in the initial value, and the equation is only mildly unstable. On the other hand, if $K > 0$, then the left-hand inequality in the theorem shows that the solution is sensitive to changes in the initial value, especially over long intervals. We can briefly summarize these results by saying that if $\partial f / \partial y < 0$ in a region of the plane, then solutions in that region get closer together as t increases, while if $\partial f / \partial y > 0$ in a region of the plane, then solutions in that region get farther apart as t increases. In particular, if $\partial f / \partial y \le 0$ everywhere, then the differential equation is stable; but if $\partial f / \partial y > 0$ everywhere, then the equation is unstable.

Remark 5.2 The right-hand inequality in the theorem is an example of a *continuous dependence* result; it shows that the solution depends continuously on the initial value.

Let us examine our examples in light of these observations. Rewriting the equation in Example 5.2 as $dy/dt = -2y + e^{-t}$, we see that $f(t, y) = -2y + e^{-t}$ and $\partial f / \partial y = -2$. We can apply the theorem with $t_0 = 0, t_1 = \infty$, and $L = K = -2$ to conclude that

$$|y(t) - \tilde{y}(t)| = |y_0 - \tilde{y}_0| e^{-2t}, \quad \text{for all } t \ge 0,$$

as we found above from the solution formula. Thus the equation is stable. We can also see that the equation is stable just by noting that $\partial f / \partial y < 0$. Similarly, for the equation of Example 5.3, $f(t, y) = 2y - 3e^{-t}$ and $\partial f / \partial y = 2 > 0$, so the equation is unstable.

We can also understand the stability of the differential equations in these two examples by examining the solution formulas. But $\partial f/\partial y$ can be calculated and its sign and size found, even if a solution formula cannot be found. For example, we can immediately tell that the differential equation $dy/dt + t^2y^3 = \cos t$ is stable because

$$f(t, y) = -t^2y^3 + \cos t$$

and $\partial f/\partial y = -3t^2y^2 \leq 0$. Yet neither **dsolve** nor any other standard technique enables us to find a formula solution to this differential equation.

Finally, we note that many equations of the form $dy/dt = f(t, y)$ cannot be classified simply as stable or unstable, because $\partial f/\partial y$ may be negative at some points and positive at others. Nonetheless, we may still be able to determine whether a particular solution is stable (insensitive to its initial value) or unstable (sensitive to its initial value) according to whether $\partial f/\partial y \leq 0$ or $\partial f/\partial y > 0$ along the solution curve. Throughout the book we will see many examples that illustrate the dependence (either sensitive or insensitive) of the solution to an initial value problem on the initial value.

5.4 Different Types of Symbolic Solutions

In Section 5.1, we showed how to evaluate and plot a symbolic solution given by **dsolve** in the ideal case that it outputs an explicit solution involving only functions built into MATLAB. As we mentioned there, many other types of output are possible, and one must deal with them differently. Here we give additional examples illustrating other possibilities.

Example 5.4 Consider the differential equation

$$\frac{dy}{dt} = t + y^2.$$

This equation is neither linear nor separable, and its solutions do not have a formula in terms of elementary functions. However, **dsolve** can still find a formula for the solutions. To get the solution with $y(0) = -3$, type:

```
>> sol2 = dsolve('Dy = t + y^2', 'y(0) = -3', 't');
>> pretty(sol2)
                   2  2/3            5/6
      (gamma(2/3)  3     + 2 pi 3    ) AiryAi(1, -t)
    - -------------------------------------------- + AiryBi(1, -t)
                        2  1/6           1/3
              -gamma(2/3)  3    + 2 pi 3
    --------------------------------------------------------------
                       2  2/3            5/6
          (gamma(2/3)  3     + 2 pi 3    ) AiryAi(-t)
        ----------------------------------------- + AiryBi(-t)
                            2  1/6           1/3
                  -gamma(2/3)  3    + 2 pi 3
```

Because the formula for the solution is so long, we used **pretty** to print it in a more readable format. This formula contains three special functions: **gamma**, **AiryAi**, and **AiryBi**. While **gamma** is a valid MATLAB function, the other two are not; this is the key difference between this example and Example 5.1.

The reason that the output of **dsolve** sometimes contains non-MATLAB functions is that **dsolve** and other commands in the Symbolic Math Toolbox are based on software licensed from another program called Maple. MATLAB attempts to convert Maple output to valid MATLAB expressions, but some Maple functions remain unconverted. In this example, the Maple functions **AiryAi(x)** and **AiryBi(x)** are not converted to their MATLAB counterparts **airy(0,x)** and **airy(2,x)**. As a result, most MATLAB commands (in particular, **ezplot**) cannot evaluate **sol2**. However, since **subs** is part of the Symbolic Math Toolbox, it can interpret Maple expressions such as **sol2**. Thus, we rely more heavily on **subs** in this case than in Example 5.1.

Typing **subs(sol2, 't', 1)** simply substitutes **1** for **t** in the formula for **sol2**, without evaluating the result numerically. You can force the output to be numeric with **double**:

```
>> double(subs(sol2, 't', 1))
ans =
   -0.4378
```

You can also define an anonymous function based on the output:

```
>> y2 = @(t) double(subs(sol2, 't', t))
```

(This construction is a bit tricky; if you type **y2(1)**, the last **t** in the definition for **y2** is replaced by **1**, but the **t** inside single quotes is not, so MATLAB evaluates the expression **double(subs(sol2, 't', 1))** as before.)

Having defined **y2**, you can plot the solution for $0 \leq t \leq 1$ by typing **ezplot(sol2, [0 1])**. Alternatively, you can use **plot**:

```
>> tvals = 0:0.01:1;
>> plot(tvals, double(subs(sol2, 't', tvals)))
```

Here is one way to plot a family of solutions. As in Example 5.1, start by defining a solution with a generic initial condition $y(0) = c$:

```
>> sol2a = dsolve('Dy = t + y^2', 'y(0) = c', 't')
```

To plot the solutions with $y(0) = -3, -2.5, -2, \ldots, 0$ on the interval $0 \leq t \leq 1$, type:

```
>> figure; hold on
>> tvals = 0:0.01:1;
>> vals = subs(sol2a, 't', tvals);
>> for cval = -3:0.5:0
      plot(tvals, double(subs(yvals, 'c', cval)))
   end
>> hold off
```

Here we first substituted the vector of t values into the solution, because these values are the same for each curve. (We did not use **double** at this point, because numerical evaluation is not necessary, nor even possible, until a value of c is specified.) The result is a vector of symbolic expressions that we called **yvals**. Then we successively substituted each of the seven values of c into **yvals**, made the result a numeric vector with **double**, and plotted this vector versus the vector of t values.

Remark 5.3 All of the commands we used in this example also work for Example 5.1, and in any other case **dsolve** yields an explicit solution to a first order equation.

Example 5.5 The differential equation

$$\frac{dy}{dt} = \frac{y^4+1}{y^5}$$

is separable, and MATLAB can do the necessary integral to solve it:

```
>> sol3 = dsolve('Dy = (y^4 + 1)/y^5')
sol3 =
t-1/2*y^2+1/2*atan(y^2)+C1 = 0
```

However, MATLAB cannot solve the resulting equation for y in terms of t, so it outputs an implicit solution. Without an explicit formula for y in terms of t, we must take a different approach to plotting and evaluating solutions.

First, we express the constant of integration, which MATLAB calls **C1**, as a function of t and y:

```
>> sol3expr = solve(sol3, 'C1')
sol3expr =
-t+1/2*y^2-1/2*atan(y^2)
```

The solution curves are the level curves of this expression, which we make into an anonymous function as follows:

```
>> sol3func = @(t, y) eval(vectorize(sol3expr))
```

Then we can use **contour** to plot several solutions, say for $0 \leq t \leq 5$ and $0 \leq y \leq 3$:

```
>> [T Y] = meshgrid(0:0.05:5, 0:0.05:3);
>> contour(T, Y, sol3func(T, Y))
>> xlabel t, ylabel y
```

The result is shown in Figure 5.5.

Unfortunately, when **dsolve** cannot find an explicit solution, it gives an error message when you try to specify an initial condition. Nonetheless, you can easily find the constant of integration for a given initial condition by plugging the appropriate values of t and y into **sol3func**. So, to plot the solution with initial condition $y(0) = 1$ on the interval $0 \leq t \leq 5$, type:

```
>> c = sol3func(0, 1);
>> [T Y] = meshgrid(0:0.05:5, 0:0.05:4);
>> contour(T, Y, sol3func(T, Y), [c c])
```

The result is shown in Figure 5.6. Remember that when making contour plots, you must choose both the horizontal and vertical ranges in the **meshgrid** command; some trial and error may be necessary to find a vertical range that is appropriate to a particular solution. To plot a family of solutions, say with initial conditions $y(0) = 1, 1.2, 1.4, \ldots, 2$, over the same range, type:

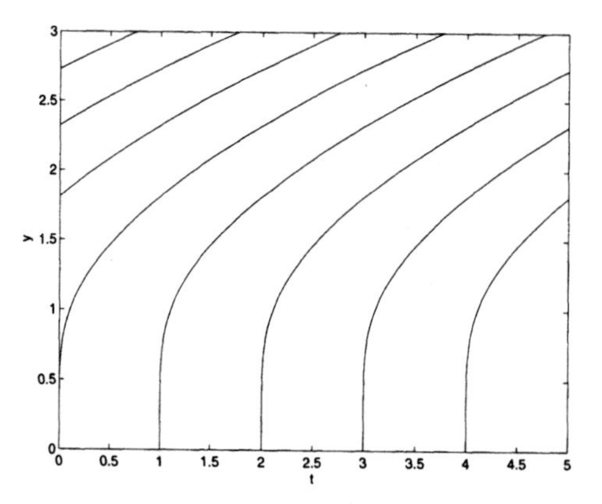

Figure 5.5: Solutions of $dy/dt = (y^4 + 1)/y^5$

```
>> cvals = sol3func(0, 1:0.2:2);
>> contour(T, Y, sol3func(T, Y), cvals)
```

To evaluate a particular solution at a particular value of t, you can substitute the appropriate value of t into **sol3func**, set it equal to the constant of integration you found from the initial condition, and solve for y. One way to do this is as follows:

```
>> syms y; solve(sol3func(4, y) - c, y)
ans =
 3.1116689668536231816288967288472
-3.1116689668536231816288967288472
```

Here **c** is the value we previously computed for the initial condition $y(0) = 1$, and we have solved for the value of $y(4)$. However, as we discussed in Section 3.7, solving algebraic equations is not always straightforward. In this case, there are two solutions to the equation for $y(4)$. Based on Figure 5.6, it is clear that the positive solution is the one we want in this case. But bear in mind that sometimes **solve** does not find all the solutions of an equation, and when it gives a single answer, that may not be the solution you want. It is safer in general to graph the solution you want and then use **fzero** to solve numerically for y using an initial guess based on the graph:

```
>> fzero(@(y) sol3func(4, y) - c, 3)
ans =
   3.1117
```

Remark 5.4 The commands in the previous paragraph illustrate some subtleties in MATLAB syntax when dealing with symbolic expressions. While it may have been more intuitive to type **solve(sol3func(4, y) = c, y)**, using an equals sign here results in an error message. Instead, we type an expression **sol3func(4, y) - c** for **solve** to set equal to zero. While **c** is already defined, **y** is undefined at this point, so we must first declare it to be a symbolic variable to avoid an error message. The function **fzero**

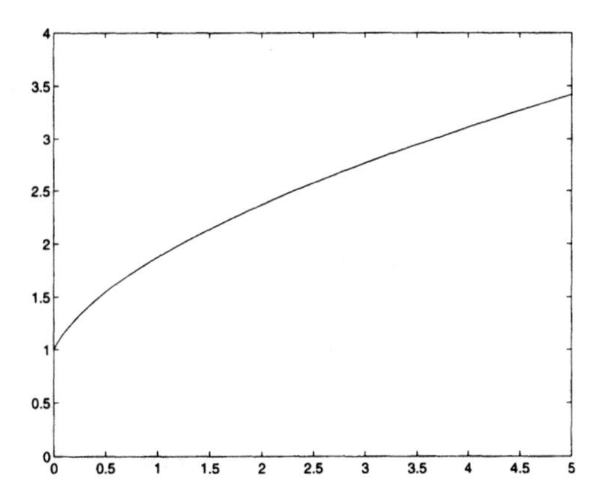

Figure 5.6: Solution of $dy/dt = (y^4 + 1)/y^5$ with $y(0) = 1$

requires its first input argument to be a function, not a symbolic expression, so we make the expression to be set to zero into an anonymous function. In this case, **y** is a dummy variable in the function definition, and need not be declared symbolic beforehand.

Example 5.6 The differential equation

$$\frac{dy}{dt} = e^{-y} + y$$

is separable, as in the previous example, but this time **dsolve** can't do the y integral:

```
>> sol4 = dsolve('Dy = exp(-y) + y', 't')
sol4 =
t-Int(1/(exp(-_a)+_a),_a =   .. y)+C1 = 0
```

In this output, **_a** is a dummy variable, and the expression **Int(1/(exp(-_a)+_a), _a =   .. y)** represents the indefinite integral $\int 1/(e^{-y} + y)dy$. You can use a formula like this in conjunction with a numerical integration routine (such as **quadl**) to obtain a highly accurate approximate solution. However, the numerical methods we will describe in Chapter 7 are generally adequate for such equations.

Remark 5.5 Although we have focused on first order equations in this section, **dsolve** also solves higher order equations (see Chapter 10) and systems of equations (see Chapter 13). The online help for **dsolve** gives examples.

There are still many differential equations that MATLAB cannot solve symbolically in any of the forms we have described. This does not mean that there is no solution, just that MATLAB cannot find a formula for the solution. As we saw in Section 5.2, solutions to $dy/dt = f(t, y)$ are guaranteed to exist for all reasonably nice functions f. You can use the methods of Chapters 6 and 7 to analyze equations for which no formula solution is available.

Finally, it is important to realize that MATLAB, like any software system, can occasionally produce misleading or incorrect results. As the following example shows, a good theoretical understanding of the nature of solutions to differential equations is a valuable guide in interpreting computer-generated results, and therefore in identifying situations in which MATLAB has produced a misleading or incorrect result.

Example 5.7 Consider the initial value problem

$$\frac{dy}{dt} = t^2 + y^2, \qquad y(0) = 1.$$

The commands

```
>> sol5 = dsolve('Dy = t^2 + y^2', 'y(0) = 1', 't');
>> ezplot(sol5, [-1 1]), title ''
```

produce the graph in Figure 5.7. The solution should be continuous, but the graph has a jump at $t = 0$. According to the initial condition, the solution should be 1 at $t = 0$, but typing **subs(sol5, 't', 0)** gives the answer **NaN**, for "not a number". The reason is that when MATLAB substitutes $t = 0$, its solution formula yields the indeterminate form ∞/∞. This would not be such a serious problem if the formula gave correct answers for t near 0, and indeed typing (for instance) **subs(sol5, 't', 0.001)** gives the answer **1.0010**. However, typing **subs(sol5, 't', -0.001)** yields **-1.0010**. This is not plausible, because for y to change from -1.001 to 1 as t changes from -0.001 to 0 would require dy/dt to be very large, whereas the differential equation requires that $dy/dt = t^2 + y^2$, which is not large in this range of t and y. We presume that **dsolve** used a solution method that assumed at some point that $t > 0$, and while its solution solves the differential equation both for $t > 0$ and $t < 0$, it is discontinuous at $t = 0$.

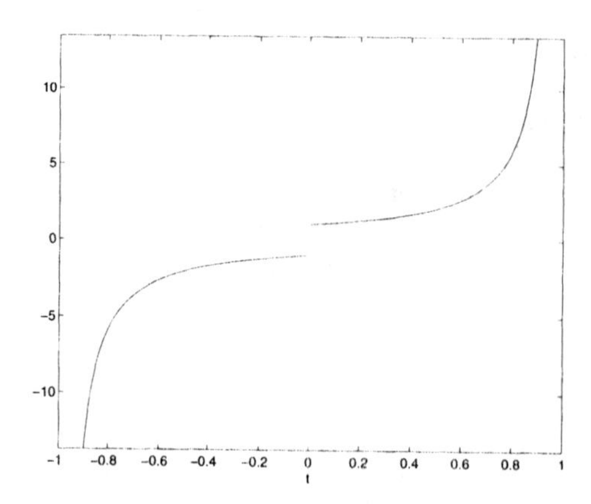

Figure 5.7: The "Solution" of $dy/dt = t^2 + y^2$, $y(0) = 1$, According to **dsolve**

Chapter 6

A Qualitative Approach to Differential Equations

In this chapter, we discuss a *qualitative approach* to the study of differential equations, and obtain qualitative information about the solutions directly from the differential equation, without the use of a solution formula.

Consider the general first order differential equation

$$\frac{dy}{dt} = f(t, y). \tag{6.1}$$

We can obtain qualitative information about the solutions $y(t)$ by viewing (6.1) geometrically. Specifically, we can obtain this information from the direction field of (6.1). Recall that the direction field is obtained by drawing through each point in the (t, y)-plane a short line segment with slope $f(t, y)$. Solutions, or integral curves, of (6.1) have the property that at each of their points they are tangent to the direction field at that point, and therefore the general *qualitative* nature of the solutions can be determined from the direction field. Direction fields can be drawn by hand for some simple differential equations, but MATLAB can draw them for any first order equation. We illustrate the qualitative approach with two examples.

6.1 Direction Field for a First Order Linear Equation

Consider the equation

$$\frac{dy}{dt} = e^{-t} - 2y. \tag{6.2}$$

MATLAB's command for plotting direction fields is **quiver**, used in conjunction with **meshgrid**. To plot the direction field of (6.2) on the rectangle $-2 \le t \le 3$, $-1 \le y \le 2$, type the following sequence of commands:

```
>> [T, Y] = meshgrid(-2:0.2:3, -1:0.2:2);
>> S = exp(-T) - 2*Y;
>> quiver(T, Y, ones(size(S)), S), axis tight
```

The first command sets up a 26 by 16 grid of uniformly spaced points in the rectangle. The second command evaluates the right-hand side of the differential equation at the grid points, thereby producing the slopes, $s = f(t, y)$, at these points. The vectors $(1, s)$ have the desired slopes but have different lengths at the different grid points. The **quiver** command, used for plotting vector fields, requires four inputs: the array **T** of t-values, the array **Y** of y-values, and arrays consisting of the two components of the direction field vectors. Since all of these arrays must have the same size, the code **ones(size(S))** conveniently creates an array of ones of the same size as **S**. In the last command, **axis tight** eliminates white space at the edges of the direction field. The result looks like this:

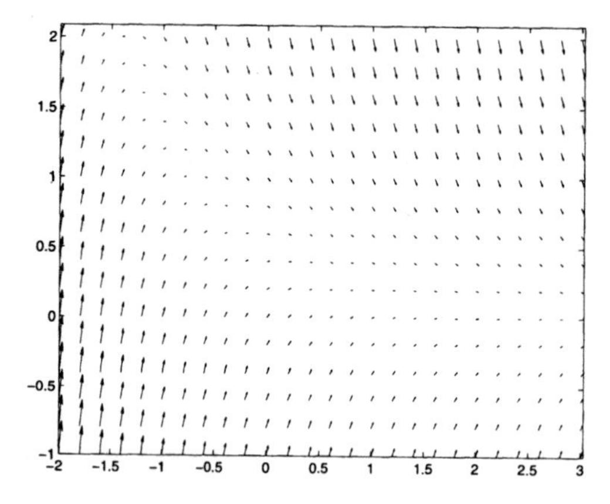

Figure 6.1: Direction Field for Equation (6.2)

While this picture is 'correct,' it is somewhat hard to read because of the fact that many of the vectors are quite small. One can get a better picture by rescaling the arrows so that they do not vary in magnitude—for example, by dividing each vector $(1, s)$ by its length $\|(1, s)\| = \sqrt{1 + s^2}$. In order to do this in MATLAB we modify the preceding sequence of commands to:

```
>> [T, Y] = meshgrid(-2:0.2:3, -1:0.2:2);
>> S = exp(-T) - 2*Y;
>> L = sqrt(1 + S.^2);
>> quiver(T, Y, 1./L, S./L, 0.5), axis tight
>> xlabel 't', ylabel 'y'
>> title 'Direction Field for dy/dt = exp(-t) - 2y'
```

The fifth entry, 0.5, in the **quiver** command reduces the length of the vectors by half and prevents the arrow heads from swallowing up the tails of nearby vectors. The result is shown in Figure 6.2.

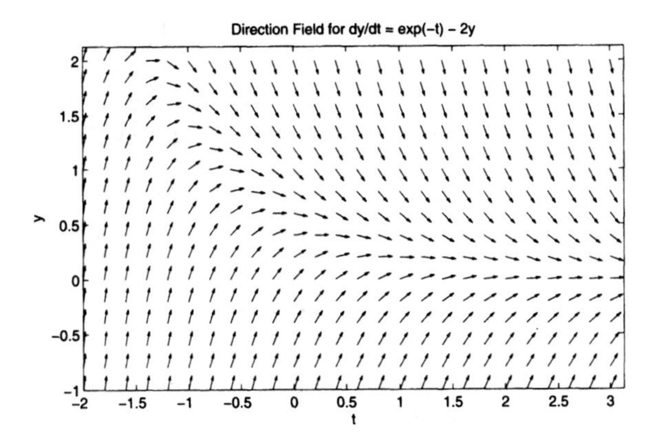

Figure 6.2: Improved Direction Field for Equation (6.2)

Note that while we're at it we have labeled the axes and the direction field plot. The direction field picture in Figure 6.2 strongly suggests that if a solution is negative at some point, then it is increasing at that point, and that all solutions approach zero as $t \to \infty$. The general solution of equation (6.2) is $e^{-t} + ce^{-2t}$.

Figure 6.3 shows several solution curves superimposed on the direction field. Note from equation (6.2) itself that $y'(t) = 0$ when $y = \frac{1}{2}e^{-t}$, and thus this curve is where the maximum points occur on the solution curves. In Figure 6.3, the curve $y = \frac{1}{2}e^{-t}$ is indicated with a dashed line.

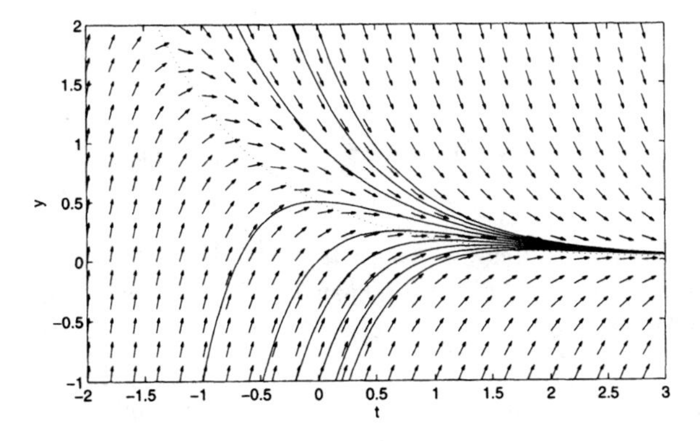

Figure 6.3: Direction Field and Solution Curves for Equation (6.2)

Exercise 6.1 Pursue this idea further by differentiating (6.2) and showing that the inflection points on the solution curves lie along the curve $y = \frac{3}{4}e^{-t}$.

6.2 Direction Field for a Non-Linear Equation

Equation (6.2) can be solved explicitly because it is linear. Now let us consider an example that cannot be solved explicitly in terms of elementary functions. (Note that **dsolve** does find an explicit solution in terms of Airy functions.)

Consider the equation

$$\frac{dy}{dt} = y^2 + t. \tag{6.3}$$

Its direction field is obtained with the MATLAB commands:

```
>> [T, Y] = meshgrid(-2:0.2:2, -2:0.2:2);
>> S = Y.^2 + T; L = sqrt(1 + S.^2);
>> quiver(T, Y, 1./L, S./L, 0.5), axis tight
>> xlabel 't', ylabel 'y'
>> title 'Direction Field for dy/dt = y^2 + t'
```

The result is shown in Figure 6.4. From the picture, it appears that all solutions are increas-

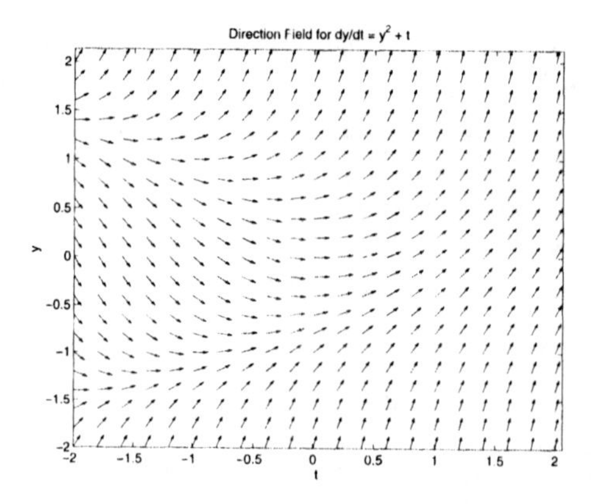

Figure 6.4: Direction Field for Equation (6.3)

ing for $t > 0$ and that all solutions eventually become positive. Figure 6.4 also suggests that all solutions approach infinity. Does this happen at a finite value of t or only as $t \to \infty$? It is impossible to decide on the basis of Figure 6.4, but in fact for each solution $y(t)$ there is a finite value t^* such that $\lim_{t \to t^*_-} y(t) = \infty$. One can see an indication of this as follows. For $t \geq 1$, the right-hand side of equation (6.3) is $\geq y^2 + 1$. But the equation

$$\frac{dy}{dt} = y^2 + 1 \tag{6.4}$$

is separable, and can be rewritten as

$$\frac{dy}{y^2 + 1} = dt$$

and thus can be solved explicitly; the solution is $\arctan y = t + C$ or $y = \tan(t + C)$. But $\tan(t + C) \to +\infty$ as $t \to \frac{\pi}{2} - C$ (from the left). Since the solutions of (6.4) blow up in finite time, then any solution of (6.3) whose domain of definition includes a value of $t \geq 1$ must blow up in finite time. By the uniqueness of solutions and a glance at the direction field in Figure 6.4, it is clear that *all* solutions must blow up in finite time.

Exercise 6.2 Use MATLAB to graph some of the direction fields in your textbook. If you can solve any of the equations explicitly, try superimposing some of the solution curves on top of the direction field.

6.3 Autonomous Equations

Equations of the form

$$\frac{dy}{dt} = f(y), \tag{6.5}$$

which do not involve t in the right-hand side, are called *autonomous* equations. If a physical system follows rules of evolution that do not change with time, then the evolution of the system is governed by an autonomous equation. Such equations are particularly amenable to qualitative analysis.

Consider equation (6.5). To be concrete, suppose $f(y)$ has two zeros y_1 and y_2, called *critical points* of the differential equation. Furthermore, suppose the graph of $f(y)$ is as shown in Figure 6.5. Then by considering the properties of $f(y)$, the direction field of (6.5)

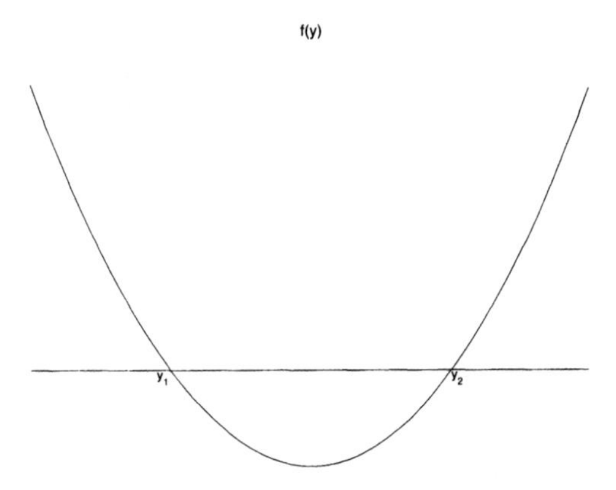

Figure 6.5: Right-hand Side of Equation (6.5)

can be easily understood and drawn by hand. At the points (t, y_1) and (t, y_2) the slope is zero; at points (t, y) with $y < y_1$, the slope is positive and increases from 0 to ∞ as y decreases from y_1 to $-\infty$; at points (t, y) with $y_1 < y < y_2$, the slope is negative, and first decreases and then increases as y increases from y_1 to y_2; at points (t, y) with $y > y_2$ the slope is positive and increases from 0 to ∞ as y increases from y_2 to ∞; finally, the slopes along any horizontal line are constant (since f does not depend on t).

The direction field thus has the general appearance shown in Figure 6.6. Note that Figure 6.5 depicts $f(y)$ vs. y, while Figure 6.6 shows y vs. t. Several facts are suggested

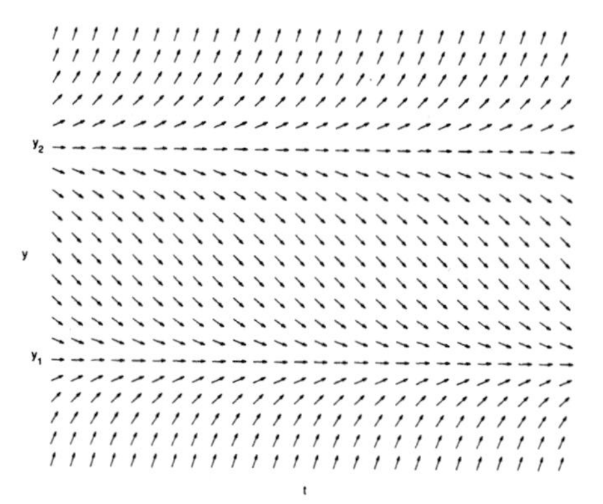

Figure 6.6: Direction Field for Equation (6.5)

by the direction field, namely, that: (i) there are two constant solutions $y = y_1$ and $y = y_2$; (ii) solutions starting above y_2 tend to ∞; and (iii) all other solutions tend to y_1 as $t \to \infty$. We can derive these properties of the solutions $y(t)$ directly from the graph of $f(y)$ as follows:

1. The constant functions $y(t) = y_1$ and $y(t) = y_2$ are solutions, called the *equilibrium solutions*. They are the unique solutions that satisfy the initial conditions $y(0) = y_1$ and $y(0) = y_2$, respectively.

2. Consider a solution $y(t)$ with $y(0) < y_1$. Then, because of the uniqueness of solutions, it must be that $y(t) < y_1$ for all t and

$$y'(t) = f(y(t)) > 0.$$

Hence $y(t)$ is an increasing function. Since it is also bounded above, it has a limit at infinity

$$\lim_{t\to\infty} y(t) = b \le y_1. \tag{6.6}$$

Could it happen that $b < y_1$? The answer is no, because if $b < y_1$, then

$$y'(t) = f(y(t)) \to f(b) > 0. \tag{6.7}$$

But equations (6.6) and (6.7) say that $y(t)$ approaches a horizontal asymptote at the same time that its slope is "permanently" bigger than a positive number. The resulting contradiction guarantees that

$$\lim_{t\to\infty} y(t) = y_1.$$

3. If $y(t)$ is a solution with $y(0) > y_2$, then $y(t) > y_2$ for all t. Also

$$y'(t) = f(y(t)) > 0.$$

Hence $y(t)$ is once again increasing. In fact, $y(t) \to \infty$. This can be shown by an argument similar to that used in part (b). Sometimes, as we shall see from the example below, $y(t)$ actually reaches ∞ in finite time.

4. Now consider $y_1 < y_0 < y_2$. Then if $y(t)$ is a solution with $y(0) = y_0$, we must have $y_1 < y(t) < y_2$ for all t. Also, $y'(t) = f(y(t)) < 0$. Hence $y(t)$ is a decreasing function. As in part 2, it is not difficult to show that $\lim_{t\to\infty} y(t) = y_1$.

Note that in this analysis the properties of $y(t)$ are determined solely from the sign of $f(y)$.

We call the critical point y_1 asymptotically stable and the critical point y_2 unstable, since solutions that start near y_1 converge to y_1, while those that start near y_2 not only do not converge to y_2, but eventually move away from y_2.

Exercise 6.3 Let $\bar{y}$ be the value between y_1 and y_2 where f takes on its minimum. Show that if $\bar{y} < y(0) < y_2$, then $y(t)$ has an inflection point where $y(t) = \bar{y}$.

6.3.1 Examples of Autonomous Equations

Now we examine a specific example, the equation

$$\frac{dy}{dt} = y^2 - y. \tag{6.8}$$

Its direction field has the general appearance of Figure 6.6, with $y_1 = 0$ and $y_2 = 1$. But we can actually derive an explicit formula solution. If $y \neq 0$ and $y \neq 1$, we can solve by separating variables:

$$\begin{aligned} t + C = \int dt &= \int \frac{dy}{y^2 - y} \\ &= \int \left(\frac{1}{y-1} - \frac{1}{y} \right) dy \\ &= \ln|y-1| - \ln|y| \\ &= \ln\left|\frac{y-1}{y}\right| \\ &= \ln\left|1 - \frac{1}{y}\right|. \end{aligned}$$

Exponentiating both sides gives

$$1 - \frac{1}{y} = ke^t,$$

where $k = \pm e^C$. Solving for y, we find that $y(t) = 1/(1 - ke^t)$. Noting that $y_0 = y(0) = 1/(1-k)$, we have

$$y(t) = \frac{y_0}{(1 - y_0)e^t + y_0}. \tag{6.9}$$

Although formula (6.9) was derived under the assumption that $y_0 \neq 0, 1$, it is easily seen to be valid for all values of y_0.

MATLAB can solve this equation and plot the solution curves. Here is a sequence of commands that does so.

```
>> figure, hold on
>> syms t c
>> sol = simplify(dsolve('Dy = y^2 - y', 'y(0) = c', 't'));
>> for cval= -1:0.25:2
    ezplot(subs(sol, c, cval), [0, 3]), end
>> xlabel 't', ylabel 'y'
>> title 'Solutions of dy/dt = y^2 - y'
```

Figure 6.7 contains the actual solution curves for the differential equation—as drawn by

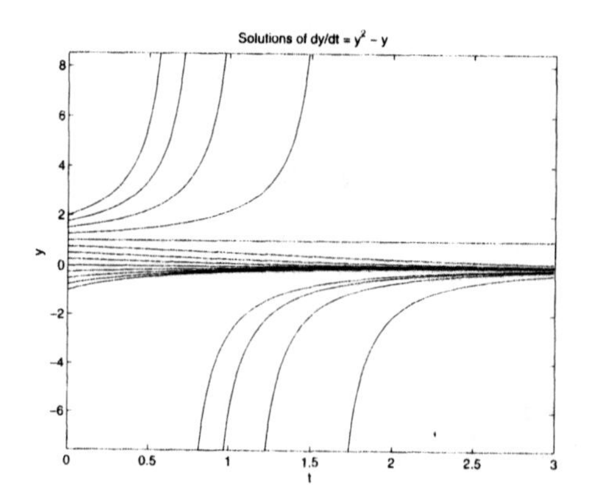

Figure 6.7: Solution Curves for Equation (6.8)

MATLAB. Note that the solution curves corresponding to $y(0) > 1$ go to infinity in finite time.

Exercise 6.4

(a) Use formula (6.9) to verify the properties of $y(t)$ obtained above by the qualitative method. Determine where the solutions are increasing or decreasing and whether there are limits as $t \to \infty$.

(b) Suppose $y_0 > 1$. Use formula (6.9) to show that $y(t) \to \infty$ in finite time. Find the time t^* at which this happens.

(c) Show that $y(t) = 0$ and $y(t) = 1$ are the equilibrium solutions. Does formula (6.9) yield these solutions?

Here is another example. The left graph in Figure 6.8 shows a function $f(y)$ with three

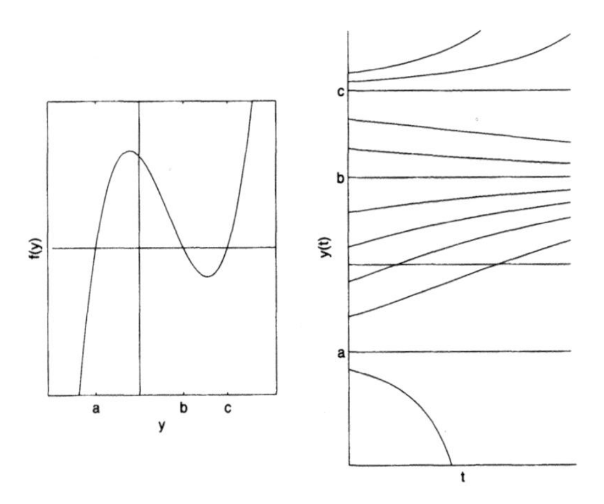

Figure 6.8: An Autonomous Equation with 3 Critical Points

zeros; thus the autonomous differential equation $y' = f(y)$ has three critical points. Let's look at the critical point in the middle. Solutions starting below b have a positive slope, so they increase toward b. Solutions starting above b have a negative slope, so they decrease toward b. Therefore, we expect b to be an asymptotically stable equilibrium solution. A similar analysis of the signs of $f(y)$ suggests that the other two critical points are unstable. The right graph in Figure 6.8 shows a few of the solution curves. As you can see, this graph illustrates the conclusions that we drew from our analysis of the graph of $f(y)$.

Problem Set B

First Order Equations

The solution to Problem 5 appears in the *Sample Solutions*.

1. Consider the initial value problem

$$ty' + 3y = 5t^2, \qquad y(2) = 5.$$

   (a) Solve using **dsolve**. Define the solution function $y(t)$ in MATLAB, and then determine its behavior as t approaches 0 from the right and as t becomes large. This can be done by plotting the solution on intervals such as $0.5 \leq t \leq 5$ and $0.2 \leq t \leq 20$.

   (b) Change the initial condition to $y(2) = 3$. Determine the behavior of this solution, again by plotting on intervals such as those mentioned in part (a).

   (c) Find a general solution of the differential equation by solving

$$ty' + 3y = 5t^2, \qquad y(2) = c.$$

   Now find the solutions corresponding to the initial conditions

$$y_j(2) = j, \qquad j = 3, \ldots, 7.$$

   Plot the functions $y_j(x)$, $j = 3, \ldots, 7$, on the same graph. Describe the behavior of these solutions for small positive t and for large t. Find the solution that is not singular at 0. Identify its plot in the graph.

2. Consider the differential equation

$$ty' + 2y = e^t. \tag{B.1}$$

With initial condition $y(1) = 1$, this has the solution

$$y(t) = \frac{e^t(t-1)+1}{t^2}.$$

(a) Verify this using MATLAB, both by direct differentiation and by using **dsolve**.

(b) Graph $y(t)$ on the interval $0 < t < 2$. Describe the behavior of the solution near $t = 0$ and for large values of t.

(c) Plot the solutions $y_j(t)$ of (B.1) corresponding to the initial conditions

$$y_j(1) = j, \qquad j = -3, -2, \ldots, 2, 3,$$

all on the same graph. Use **axis** to adjust the picture.

(d) What do the solutions have in common near $t = 0$? for large values of t? Is there a solution to the differential equation that has no singularity at $t = 0$? If so, what is it?

3. Consider the initial value problem

$$ty' + y = 2t, \qquad y(1) = c.$$

(a) Solve it using MATLAB.

(b) Evaluate the solution with $c = 0.8$ at $t = 0.01, 0.1, 1, 10$. Do the same for the solutions with $c = 1$ and $c = 1.2$.

(c) Plot the solutions with $c = 0.8, 0.9, 1.0, 1.1, 1.2$ together on the interval $(0, 2.5)$.

(d) How do changes in the initial data affect the solution as $t \to \infty$? as $t \to 0^+$?

4. Solve the initial value problem

$$y' - 2y = \sin 2t, \qquad y(0) = c.$$

Use MATLAB to graph solutions for $c = -0.5, -0.45, \ldots, -0.05, 0$. Display all the solutions on the same interval between $t = 0$ and an appropriately chosen right endpoint. Explain what happens to the solution curves as t increases. You should identify three distinct types of behavior. Which values of c correspond to which behaviors? Now, based on this problem, and the material in Chapters 5 and 6, discuss what effect small changes in initial data can have on the global behavior of solution curves.

5. Consider the differential equation

$$\frac{dy}{dt} = \frac{t - e^{-t}}{y + e^y}$$

(*cf.* Problem 7, Section 2.2 in Boyce & DiPrima).

(a) Solve it using **dsolve**. Observe that the solution is given implicitly. Express it clearly in the form

$$f(t, y) = c.$$

(b) Use **contour** to see what the solution curves look like. For your t and y ranges, you might use **-1:0.05:3** and **-2:0.05:2**. Plot 30 contours.

(c) Plot the solution satisfying the initial condition $y(1.5) = 0.5$.

(d) To find a numerical value for the solution $y(t)$ from part (c) at a particular value of t, you can plug the value t into the equation

$$f(t, y) = f(1.5, 0.5)$$

and solve for y. Because there may be multiple solutions, you should look for one near $y = 0.5$. This can be done using **fzero**. (See Section 3.7.) Find $y(0), y(1), y(1.8), y(2.1)$. Mark these values on your plot.

6. Consider the differential equation

$$e^y + (te^y - \sin y)\frac{dy}{dt} = 0.$$

(a) Solve using **dsolve**. Observe that the solution is given implicitly in the form

$$f(t, y) = c.$$

(b) Use **contour** (see Section 3.8.6) to see what the solution curves look like. For your t and y ranges, you might use **-1:0.1:4** and **0:0.1:3**. Plot 30 contours.

(c) Plot the solution satisfying the initial condition $y(2) = 1.5$.

(d) Find $y(1), y(1.5), y(3)$. Mark these values on your plot. (See Problem 5, part (d) for suggestions.)

7. Consider the differential equation

$$y' = \frac{t^2}{1 + y^2}$$

(*cf.* Problem 8, Section 2.2 in Boyce & DiPrima).

(a) Solve it using **dsolve**. Observe that in some sense MATLAB is "too good" in that it finds three rather complicated explicit solutions. Note that two of them are complex-valued. Select the real one. In fact, for this solution it is easier to work with an implicit form; so here are instructions for converting your solution. Substitute $u = t^3$ (or $t = u^{(1/3)}$) to generate an expression in u. Set it equal to y and **solve** for u. Replace u by t^3 to get an implicit solution of the form

$$f(t, y) = c.$$

(b) Use **contour** to see what the solution curves look like. For both your t and y ranges, you might use **-1.5:0.05:1.5**. Plot 30 contours.

(c) Plot the solution satisfying the initial condition $y(0.5) = 1$.

(d) To find a numerical value for the solution $y(t)$ from part (c) at a particular value of t, you can plug the value t into the equation

$$f(t, y) = f(0.5, 1)$$

and solve for y. This can be done using **fzero**. (See Section 3.7.) Find $y(-1), y(0), y(1)$. Mark these values on your plot.

8. In this problem, we study continuous dependence of solutions on initial data.

(a) Solve the initial value problem

$$y' = y/(1 + t^2), \qquad y(0) = c.$$

(b) Let y_c denote the solution in part (a). Use MATLAB to plot the solutions y_c for $c = -10, -9, \ldots, -1, 0, 1, \ldots, 10$ on one graph. Display all the solutions on the interval $-20 \le t \le 20$.

(c) Compute $\lim_{t \to \pm\infty} y_1(t)$.

(d) Now find a single constant M such that for all real t, we have

$$|y_a(t) - y_b(t)| \le M|a - b|,$$

for any pair of numbers a and b. Show from the solution formula that $M = 1$ will work if we consider only negative values of t.

(e) Relate the fact in (d) to Theorem 5.2.

9. Use **dsolve** to solve the following differential equations or initial value problems from Boyce & DiPrima. In some cases, MATLAB will not be able to solve the equation. (Before moving on to the next equation, make sure you haven't mistyped something.) In other cases, MATLAB may give extraneous solutions. (Sometimes these correspond to non-real roots of the equation it solves to get y in terms of t.) If so, you should indicate which solution or solutions are valid. You also might try entering alternative forms of an equation, for example, $M + Ny' = 0$ instead of $y' = -M/N$, or vice versa.

(a) $y' = ry - ky^2$ (Sect. 2.4, Prob. 29),

(b) $y' = t(t^2 + 1)/(4y^3)$, $y(0) = -1/\sqrt{2}$ (Sect. 2.2, Prob. 16),

(c) $(e^t \sin y + 3y)\, dt - (3t - e^t \sin y)\, dy = 0$ (Sect. 2.6, Prob. 8),

(d) $\dfrac{dy}{dt} = (2y - t)/(2t - y)$ (Sect. 2.2, Prob. 30),

(e) $\dfrac{dy}{dt} = (2t + y)/(3 + 3y^2 - t)$, $y(0) = 0$ (Ch. 2, Miscellaneous Problems, Prob. 3).

10. Use **dsolve** to solve the following differential equations or initial value problems from Boyce & DiPrima. See Problem 9 for additional instructions.

    (a) $t^3 y' + 4t^2 y = e^{-t}$, $y(-1) = 0$ (Sect. 2.1, Prob. 19),

    (b) $y' + (1/t)y = 3\cos 2t$, $t > 0$ (Sect. 2.1, Prob. 40),

    (c) $y' = ty(4 - y)/3$, $y(0) = y_0$, $t > 0$ (Sect. 2.2, Prob. 27),

    (d) $(\frac{y}{t} + 6t)\,dt + (\ln t - 2)\,dy = 0$, $t > 0$ (Sect. 2.6, Prob. 10),

    (e) $y' = \dfrac{t^2 + ty + y^2}{t^2}$ (Sect. 2.2, Prob. 31),

    (f) $ty' + ty = 1 - y$, $y(1) = 0$ (Ch. 2, Miscellaneous Problems, Prob. 6).

11. Chapter 6 describes how to plot the direction field for a first order differential equation. For each equation below, plot the direction field on a rectangle large enough (but not too large) to show clearly all of its equilibrium points. Find the equilibria and state whether each is stable or unstable. If you cannot determine the precise value of an equilibrium point from the equation or the direction field, use **fzero** or **solve** as appropriate.

    (a) $y' = -y(y - 2)(y - 4)/10$,

    (b) $y' = y^2 - 3y + 1$,

    (c) $y' = 0.1y - \sin y$.

12. In this problem, we use the direction field capabilities of MATLAB to study two nonlinear equations, one autonomous and one non-autonomous.

    (a) Plot the direction field for the equation

    $$\frac{dy}{dt} = 3\sin y + y - 2$$

    on a rectangle large enough (but not too large) to show all possible limiting behaviors of solutions as $t \to \infty$. Find approximate values for all the equilibria of the system (you should be able to do this with **fzero** using guesses based on the direction field picture), and state whether each is stable or unstable.

    (b) Plot the direction field for the equation

    $$\frac{dy}{dt} = y^2 - ty,$$

    again using a rectangle large enough to show the possible limiting behaviors. Identify the unique constant solution. Why is this solution evident from the differential equation? If a solution curve is ever below the constant solution, what must its limiting behavior be as t increases? For solutions lying above the constant solution, describe two possible limiting behaviors as t increases. There is a solution curve that lies along the boundary of the two limiting behaviors. What does it do as t increases? Explain (from the differential equation) why no other limiting behavior is possible.

(c) Confirm your analysis by using **dsolve** on the system $y' = y^2 - ty$, $y(0) = c$, and then examining different values of c.

13. The solution of the differential equation

$$y' = \frac{2y - t}{2t - y}$$

is given implicitly by $|t - y| = c|t + y|^3$. (This is not what **dsolve** produces, which is a more complicated explicit solution for y in terms of t, but it's what you get by making the substitution $v = y/t$, $y = tv$, $y' = tv' + v$ and separating variables. You do not need to check this answer.) However, it is difficult to understand the solutions directly from this algebraic information.

(a) Use **quiver** and **meshgrid** to plot the direction field of the differential equation.

(b) Use **contour** to plot the solutions with initial conditions $y(2) = 1$ and $y(0) = -3$. (Note that **abs** is the absolute value function in MATLAB.) Use **hold on** to put these plots and the vector field plot together on the same graph.

(c) For the two different initial conditions in part (b), use your pictures to estimate the largest interval on which the unique solution function is defined.

14. Consider the differential equation

$$y' = -ty^3.$$

(a) Use MATLAB to plot the direction field of the differential equation. Is there a constant solution? If $y(0) > 0$, what happens as t increases? If $y(0) < 0$, what happens as t increases?

(b) The direction field suggests that the solution curves are symmetric with respect to the y-axis, that is, if $y_1(t)$ is a solution to the differential equation, then so is $-y_1(t)$. Verify that this is so directly from the differential equation.

(c) Use **dsolve** to solve the differential equation, thereby obtaining a solution of the form

$$y(t) = \frac{\pm 1}{\sqrt{t^2 + c}}.$$

Note that $y(0) = \pm 1/\sqrt{c}$.

Let's consider the curves above the y-axis. The ones below it are their reflections. (What about the y-axis itself?) In graphing $y(t)$, it helps to consider three separate cases: (i) $c > 0$, (ii) $c = 0$, and (iii) $c < 0$.

(d) In each case, graph the solution for several specific values of c, and identify important features of the curves. In particular, in case (iii), compute the location of the asymptotes in terms of c.

(e) Identify five different types of solution curves for the differential equation that lie above the y-axis. In each case, specify the t-interval of existence, whether the solution is increasing or decreasing, and the limiting (asymptotic) behavior at the ends of the interval.

(f) Combine the direction field plotted in (a) with the graphs plotted in (d).

15. Use **dsolve** to solve the differential equation

$$y' = ty^3 \tag{B.2}$$

with the initial condition $y(0) = c$.

(a) How many solutions does MATLAB give? Do they all satisfy the initial condition? Why do you think that MATLAB gives multiple answers?

(b) What happens if you substitute $c = 0$ in the solution to (B.2)? What happens if you instead take the limit of the solution as $c \to 0$? Do you indeed get the correct solution of the equation for this case?

(c) Plot the solutions to (B.2) for $c = -5, -4, \ldots, 4, 5$ on the same set of axes. Is the initial value problem stable?

16. Consider the critical threshold model for population growth

$$y' = -(2 - y)y.$$

(a) Find the equilibrium solutions of the differential equation. Now draw the direction field, and use it to decide which equilibrium solutions are stable and which are unstable. In particular, what is the limiting behavior of the solution if the initial population is between 0 and 2? greater than 2?

(b) Use **dsolve** to find the solutions with initial values $1.5, 0.3$, and 2.1, and plot each of these solutions. Find the inflection point for the first of these solutions.

(c) Plot the three solutions together with the direction field on the same graph. Do the solutions follow the direction field as you expect it to?

17. Consider the following logistic-with-threshold model for population growth:

$$y' = y(1 - y)(y - 3).$$

(a) Find the equilibrium solutions of the differential equation. Now draw the direction field, and use it to decide which equilibrium solutions are stable and which are unstable. In particular, what is the limiting behavior of the solution if the initial population is between 1 and 3? greater than 3? exactly 1? between 0 and 1?

(b) Next replace the logistic law by the Gompertz model, but retain the threshold feature. The equation becomes

$$y' = y(1 - \ln y)(y - 3).$$

Once again, find the equilibrium solutions and draw the direction field. You will have difficulty "reading the field" between 2.5 and 3. There appears to be a continuum of equilibrium solutions.

(c) Plot the function $f(y) = y(1 - \ln y)(y - 3)$ on the interval $0 \le y \le 4$, and then use **limit** to evaluate $\lim_{y \to 0} f(y)$.

(d) Use these plots and the discussion in Chapter 6 to decide which equilibrium solutions are stable and which are unstable. Now use the last plot to explain why the direction field (for $2.5 \le y \le 3$) appears so inconclusive regarding the stability of the equilibrium solutions. (*Hint*: The maximum value of f is a relevant number.)

18. This problem is based on Example 1 in Section 2.3 of Boyce & DiPrima: "A tank contains Q_0 lb of salt dissolved in 100 gal of water. Water containing 1/4 lb of salt per gallon enters the tank at a rate of 3 gal/min, and the well-stirred solution leaves the tank at the same rate. Find an expression for the amount of salt $Q(t)$ in the tank at time t."

The differential equation

$$Q'(t) = 0.75 - 0.03Q(t)$$

models the problem (*cf.* equation (2) in Section 2.3 of Boyce & DiPrima).

(a) Plot the right-hand side of the differential equation as a function of Q, and identify the critical point.

(b) Analyze the long-term behavior of the solution curves by examining the sign of the right-hand side of the differential equation, in a similar fashion to the discussion in Section 6.3.

(c) Use MATLAB to plot the direction field of the differential equation. In choosing the rectangle for the direction field be sure to include the point $(0, 0)$ and the critical value of Q.

(d) Use the direction field to estimate the limiting amount of salt and to determine how the amount of salt approaches this limit.

(e) Use **dsolve** to find the solution $Q(t)$ and plot it for several specific values of Q_0. Do the solutions behave as indicated in parts (b) and (d)? You should combine the direction field plot from (c) with that of the solution curves.

19. A 10-gallon tank contains a mixture consisting of 1 gallon of water and an undetermined number $S(0)$ of pounds of salt in the solution. Water containing 1 lb/gal

of salt begins flowing into the tank at the rate of 2 gal/min. The well-mixed solution flows out at a rate of 1 gal/min. Derive the differential equation for $S(t)$, the number of pounds of salt in the tank after t minutes, that models this physical situation. (*Note*: At time $t = 0$ there is 1 gallon of solution, but the volume increases with time.) Now draw the direction field of the differential equation on the rectangle $0 \leq t \leq 10$, $0 \leq S \leq 10$. From your plot,

(a) find the value A of $S(0)$ below which the amount of salt is a constantly increasing function, but above which the amount of salt will temporarily decrease before increasing;

(b) indicate how the nature of the solution function in case $S(0) = 1$ differs from all other solutions.

Now use **dsolve** to solve the differential equation. Reinforce your conclusions above by

(c) algebraically computing the value of A;

(d) giving the formula for the solution function when $S(0) = 1$;

(e) giving the amount of salt in the tank (in terms of $S(0)$) when it is at the point of overflowing;

(f) computing, for $S(0) > A$, the minimum amount of salt in the tank, and the time it occurs;

(g) explaining what principle guarantees the truth of the following statement: If two solutions S_1, S_2 correspond to initial data $S_1(0)$, $S_2(0)$ with $S_1(0) < S_2(0)$, then for any $t \geq 0$, it must be that $S_1(t) < S_2(t)$.

20. In this problem, we use **dsolve** and **solve** to model some population data. The procedure will be:

(i) Assume a model differential equation involving unknown parameters.

(ii) Use **dsolve** to solve the differential equation in terms of the parameters.

(iii) Use **solve** to find the values of the parameters that fit the given data.

(iv) Make predictions based on the results of the previous steps.

(a) Let's use the model

$$\frac{dp}{dt} = ap + b, \quad p(0) = c,$$

where p represents the population at time t. Check to see that **dsolve** can solve this initial value problem in terms of the unknown constants a, b, c. Then define a function that expresses the solution at time t in terms of a, b, c, and t. Give physical interpretations to the constants a, b and c.

(b) Next, let's try to model the population of Nevada,which was one of the fastest growing states in the U.S. during the second half of the twentieth century. Here is a table of census data:

Year	Population in thousands
1950	160.1
1960	285.3
1970	488.7
1980	800.5
1990	1201.8

We would like to find the values of a, b, and c that fit the data. However, with three unknown constants we will not be able to fit five data points. Use **solve** to find the values of a, b, and c that give the correct population for the years 1960, 1970, and 1980. We will later use the data from 1950 and 1990 to check the accuracy of the model. *Important*: In this part let t represent the time in years since 1950, because MATLAB may get stuck if you ask it to fit the data at such high values of t as 1960–1980.

(c) Now define a function of t that expresses the predicted population in year t using the values of a, b, c found in part (b). Find the population this model gives for 1950 and 1990, and compare with the values in the table above. Use the model to predict the population of Nevada in the year 2000, and to predict when the population will reach 3 million. How would you adjust these predictions based on the 1950 and 1990 data? What adjustment to a and/or b might you make? Finally, graph the population function that the model gives from 1950 to 2050, and describe the predicted future of the population of Nevada, including the limiting population (if any) as $t \to \infty$.

21. Consider $y' = (\alpha - 1)y - y^3$.

(a) Use **solve** to find the roots of $(\alpha - 1)y - y^3$. Explain why $y = 0$ is the only real root when $\alpha \leq 1$, and why there are three distinct real roots when $\alpha > 1$.

(b) For $\alpha = -2, -1, 0$, draw a direction field for the differential equation and deduce that there is only one equilibrium solution. What is it? Is it stable?

(c) Do the same for $\alpha = 1$.

(d) For $\alpha = 1.5, 2$, draw the direction field. Identify all equilibrium solutions, and describe their stability.

(e) Explain the following statement: "As α increases through 1, the stable solution $x = 0$ *bifurcates* into two stable solutions."

22. In Chapter 6, we discussed the equation (6.3) and the fact that its solutions blow up in finite time. In this problem, we explore the solutions in a little more detail.

(a) Use **dsolve** to solve the initial value problem $y' = y^2 + t$ with the initial condition $y(0) = 0$. You may find that MATLAB writes the solution in terms of some mysterious functions that are unfamiliar to you, namely **AiryAi** and **AiryBi**. You can find out more about these by typing **mhelp AiryAi**, at least if you are not using the Student Version of MATLAB. (The command **mhelp** retrieves help about commands that are internal to the Maple kernel, used for symbolic computations. Unfortunately, **mhelp** is not available in the Student Version.) In fact, **AiryAi** and **AiryBi** are not defined in MATLAB itself, which makes working with them a bit complicated.

(b) Plot the solution you have obtained over the interval $-1 \leq t \leq 2$. You will find that **ezplot** fails in this case, because of the difficulty with **AiryAi** and **AiryBi** just explained in (a). However, you can plot the solution by defining a vector **T** of values of t, letting **Y = subs(sol, t, T)**, and using **plot**. You may have to experiment with the spacing of the values of t and with the scale on the y-axis to get a good picture. At what value t^* of t does it seem that the solution "blows up"? What happens to your graph past this value of t^*? Do you think it has any validity? Why or why not?

(c) Compare your estimated value of t^* with the upper bound for t^* obtained in Chapter 6 by comparison with the equation $y' = y^2 + 1$ for $t \geq 1$. Note that to make the comparison, you need to use the solution $y = \tan(t + C)$ of $y' = y^2 + 1$ that matches your solution to $y' = y^2 + t$ at $t = 1$, so you will need to estimate the appropriate value of C as well.

(d) Superimpose your graph of the solution to (6.3) on top of the direction field for the equation, and visually verify the tangency of the solution curves to the direction field.

Chapter 7

Numerical Methods

In this book you have seen that many differential equations can be solved explicitly in terms of the elementary functions of calculus. We have seen in Chapter 5 that **dsolve** solves many equations in terms of these functions.

But many other differential equations cannot be solved explicitly in terms of the functions of calculus. Consider, for example, the equation

$$\frac{dy}{dt} = e^{-t^2}.$$

Its solutions are the integrals, or antiderivatives, of e^{-t^2},

$$y(t) = \int e^{-t^2}\,dt + C,$$

but it is known that these integrals cannot be expressed in terms of the functions of calculus. (MATLAB does write this integral in closed form in terms of a function **erf**, but this function, called the "error function," is *defined* by the formula $\operatorname{erf}(x) = \frac{2}{\sqrt{\pi}}\int_0^x e^{-t^2}\,dt$, and cannot be rewritten in terms of exponentials, logs, trigonometric functions, *etc*. This is the simplest example of a *special function* as discussed in the beginning of Chapter 5.)

We will refer to solutions that can be explicitly written in term of elementary functions or special functions as *formula* or *symbolic solutions*—whether they were obtained via hand-calculation or via **dsolve**. When we cannot solve a differential equation in this way, or if the formula we find is too complicated, we turn to numerical methods to solve an initial value problem. This is similar to a situation in calculus: if we cannot find an antiderivative in terms of elementary functions, we turn to a numerical method such as the trapezoidal rule or Simpson's rule to evaluate a definite integral.

With numerical methods as with qualitative methods, we obtain information about the solution—quantitative in the one case and qualitative in the other—without the use of a solution formula.

7.1 Numerical Solutions Using MATLAB

Suppose we are interested in finding the solution to the initial value problem

$$\frac{dy}{dt} = f(t, y), \qquad y(t_0) = y_0$$

on an interval $a \le t \le b$ containing t_0, and suppose that we do not have a formula for $y(t)$. In such a situation, our strategy will be to produce a function $y_a(t)$ that both is a good approximation to $y(t)$ and can be calculated for any $t \in [a, b]$. The subscript a on y_a stands for *approximate solution*. Such a function $y_a(t)$ can be found with **ode45**, MATLAB's primary numerical differential equation solver.

We illustrate the use of **ode45** by considering the initial value problem

$$\frac{dy}{dt} = \frac{t}{y}, \qquad y(0) = 1. \tag{7.1}$$

Its exact solution is easily found to be

$$y(t) = \sqrt{t^2 + 1}.$$

Since we have an explicit formula for this solution, we will be able to compare the approximate solution $y_a(t)$ and the exact solution $y(t)$.

The first step is to define a MATLAB function equal to the right-hand side of the differential equation. There are two ways to do this. One way is to define an *anonymous function*, as in Chapter 3, by typing:

```
>> f = @(t, y) t./y
```

(Note the dot in front of the division sign; the function must be able to operate on vectors.) The other way is via a function M-file, as in Chapter 4. For our problem, the M-file could, for example, be called `func.m`, and would be:

```
function z = func(t, y)
z = t./y;
```

We recommend that you use anonymous functions whenever possible. However, M-files allow you to use more than one line to define the function, which may be necessary if the right-hand side is sufficiently complicated.

To calculate and plot the approximate solution $y_a(t)$ on the interval $[0, 2]$, type

```
>> ode45(f, [0 2], 1)
```

if **f** is an anonymous function, or

```
>> ode45(@func, [0 2], 1)
```

if `func.m` is an M-file. There is still another possibility, namely to put the definition of the anonymous function directly into the **ode45** command, like this:

```
>> ode45(@(t, y) t./y, [0 2], 1)
```

From here on we will assume that you have defined the right-hand side of the differential equation as an anonymous function **f**. The second argument to **ode45** specifies the interval on which to solve, and the third argument gives the initial value of the solution at the first point in this interval. The result is shown in Figure 7.1. The circles indicate the points at which **ode45** computed the approximate solution; the centers of these circles are connected by line segments.

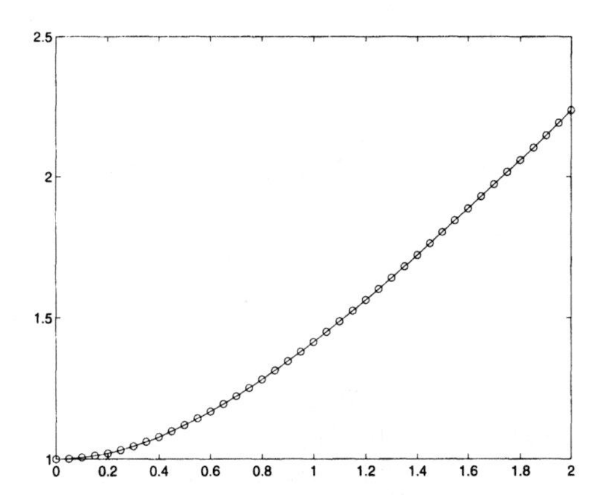

Figure 7.1: Numerical solution of $dy/dt = t/y$, $y(1) = 1$

Exercise 7.1 Display both $y_a(t)$ and the exact solution $y(t) = \sqrt{t^2 + 1}$ on the same graph. Can you distinguish between the two curves?

You can also plot a family of approximate solutions by using a vector of initial values as the third argument to **ode45**. For example, to plot the solutions of $dy/dt = t/y$ with $y(0) = 1, 1.2, 1.4, \ldots, 3$, type

```
>> ode45(f, [0 2], 1:0.2:3)
```

This procedure for plotting a family of solutions works as long as none of the solutions has a singularity in the specified interval; *cf.* the initial value problem (7.2) below and the following Exercise. See Section 8.6.2 for additional information on this point.

By default, **ode45** produces a plot of an approximate solution. To extract numerical values type

```
>> [t, ya] = ode45(f, [0 2], 1);
```

The t values at which **ode45** computes the approximate solution are stored in the vector **t**, and the corresponding approximate solution values are stored in the vector **ya**. The semicolon suppresses printing of the vectors. When used this way, **ode45** does not produce a graph; to obtain a graph, type **plot(t, ya)**. Note that **plot(t, ya)** produces a graph without circles.

To display the values of t and $y_a(t)$ in columns side-by-side, type **[t ya]**. Notice that the t values are chosen by **ode45** and may include more values, or different values, than you intended. If you are interested only in the value of $y_a(t)$ at the end of the interval, in this case $t = 2$, you can type **ya(end)**, which prints the last element in the vector **ya**.

To specify the t values, use a vector (with more than two elements) as the second argument to **ode45**. For example, to compute the approximate solution at $t = 0, 0.2, 0.4, \ldots, 2$, type

```
>> [t, ya] = ode45(f, 0:0.2:2, 1);
```

Then to make a table of the values of t and $y_a(t)$, along with the corresponding values of the exact solution $y(t) = \sqrt{t^2 + 1}$, type

```
>> y = sqrt(t.^2 + 1);
>> format long
>> [t ya y]
```

We used **format long** to obtain 15-digit output in order to see the difference between **ya** and **y**. The result is shown in Table 7.1. The approximate solution is correct to at

t	$y_a(t)$	$y(t)$
0	1.00000000000000	1.00000000000000
0.20000000000000	1.01980390030003	1.01980390271856
0.40000000000000	1.07703295800176	1.07703296142690
0.60000000000000	1.16619037435052	1.16619037896906
0.80000000000000	1.28062484165690	1.28062484748657
1.00000000000000	1.41421355599242	1.41421356237310
1.20000000000000	1.56204992884201	1.56204993518133
1.40000000000000	1.72046504740386	1.72046505340853
1.60000000000000	1.88679622083269	1.88679622641132
1.80000000000000	2.05912602304385	2.05912602819740
2.00000000000000	2.23606797273725	2.23606797749979

Table 7.1: Approximate and Exact Solutions of $dy/dt = t/y$

least eight digits in all cases. The **ode45** command attempts to produce an approximate solution with error less than or equal to 10^{-3}, though sometimes the accuracy may be better or worse. See Sections 7.3 and 7.4 for additional information on accuracy.

The **ode45** command will only compute a solution on an interval with the initial condition given at one of the endpoints, t_0. We have already described how to use **ode45** on an interval of the form $[t_0, b]$; to use it on an interval of the form $[a, t_0]$, enter the values of t in decreasing order. For example, to plot an approximate solution of the initial value problem (7.1) on the interval $[-1, 0]$, type

```
>> ode45(f, [0 -1], 1)
```

To specify values for t, use a negative increment, as in **0:-0.2:-1**.

Now suppose you want to plot the family of solutions of $dy/dt = t/y$ with initial values $y(0) = 1, 1.2, 1.4, \ldots, 3$ on the interval $[-1, 2]$. Using **ode45** with **[-1 2]** as the second argument would produce an incorrect graph, because the initial conditions would be applied at $t = -1$ instead of $t = 0$. Here is a series of commands that produces the correct graph by computing the approximate solutions separately on the intervals $[-1, 0]$ and $[0, 2]$:

```
>> [tbak, ybak] = ode45(f, [0 -1], 1:0.2:3);
>> plot(tbak, ybak)
```

```
>> hold on
>> [tfor, yfor] = ode45(f, [0 2], 1:0.2:3);
>> plot(tfor, yfor)
>> hold off
```

The result is shown in Figure 7.2.

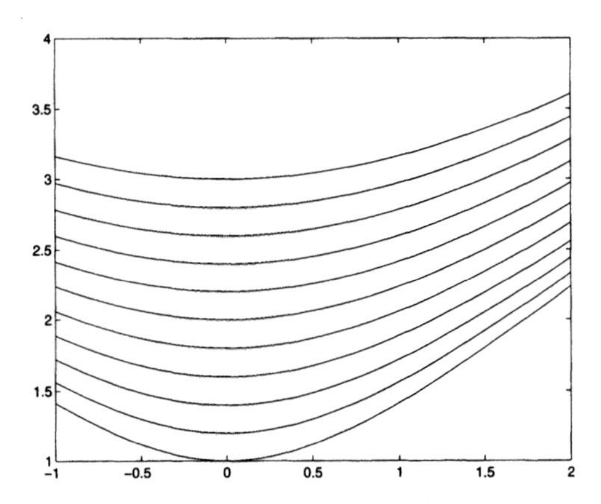

Figure 7.2: A Family of Numerical Solutions

We have illustrated **ode45** with the initial value problem (7.1), which can easily be solved explicitly. We can just as readily apply **ode45** to any first order initial value problem. Consider, for example,

$$\frac{dy}{dt} = t + y^2, \qquad y(0) = 1, \tag{7.2}$$

which cannot be solved in terms of elementary functions. The commands

```
>> g = @(t, y) t + y.^2;
>> ode45(g, [0 1], 1);
>> axis([0 1 0 100])
```

will graph an approximate solution.

Exercise 7.2 The solution $y(t)$ to (7.2) is increasing for t positive, and there is a finite value t^* such that $\lim_{t \to t^*} y(t) = \infty$; the interval of existence of the solution is $(-\infty, t^*)$. Execute the previous three commands. Does **ode45** reveal the value of t^*?

7.2 Some Numerical Methods

In order to give some idea about how **ode45** calculates approximate values, we are going to discuss several numerical methods. We will apply each method to problem (7.1).

We want to approximate the solution of the initial value problem

$$\frac{dy}{dt} = f(t, y), \qquad y(t_0) = y_0$$

on an interval $t_0 \leq t \leq b$. For n a positive integer, we divide the interval into n parts using points

$$t_0 < t_1 < t_2 < \cdots < t_n = b.$$

For simplicity, we assume that each part has the same width, or *step size*, $h = t_{i+1} - t_i = (b - t_0)/n$; therefore $t_i = t_0 + ih$. At each point t_i we seek an approximation, which we call y_i, to the true solution $y(t_i)$ at t_i:

$$y(t_i) \approx y_i.$$

7.2.1 The Euler Method

The simplest numerical solution method is due to Euler and is based on the tangent line approximation to a function. Given the initial value y_0, we define y_i recursively by

$$y_{i+1} = y_i + h\, f(t_i, y_i), \qquad i = 0, 1, \ldots, n - 1.$$

This formula is derived as follows:

$$\begin{aligned} y(t_{i+1}) &\approx y(t_i) + hy'(t_i), && \text{by the tangent line approximation} \\ &= y(t_i) + hf(t_i, y(t_i)), && \text{using the differential equation} \\ &\approx y_i + hf(t_i, y_i), && \text{since } y(t_i) \approx y_i \\ &= y_{i+1}. \end{aligned}$$

The approximations $y(t_i) \approx y_i$ should become better and better as h is taken smaller and smaller.

Example 7.1 Consider the initial value problem (7.1):

$$\frac{dy}{dt} = \frac{t}{y}, \qquad y(0) = 1.$$

We wish to approximate $y(0.3)$ using the Euler Method with step size $h = 0.1$ and three steps. We find

$$\begin{aligned} t_0 &= 0, \quad t_1 = 0.1, \quad t_2 = 0.2, \quad t_3 = 0.3 \\ y_0 &= 1 \\ y_1 &= y_0 + hf(t_0, y_0) = y_0 + ht_0/y_0 = 1 \\ y_2 &= y_1 + ht_1/y_1 = 1.01 \\ y_3 &= 1.0298. \end{aligned}$$

Since $y(0.3) = \sqrt{1.09} = 1.044030...$, we find that

$$\text{Error} = |y(0.3) - y_3| = 0.0142....$$

Next use $h = 0.05$ and 6 steps:

$$\begin{aligned} y_0 &= 1 \\ y_1 &= 1 \\ y_2 &= 1.0025 \\ y_3 &= 1.0075 \\ y_4 &= 1.0149 \\ y_5 &= 1.0248 \\ y_6 &= 1.0370 \end{aligned}$$

$$\text{Error} = |y(0.3) - y_6| = 0.0070\ldots.$$

If the initial value problem is "sufficiently smooth", then for the Euler Method one can show that the error in stepping from t_0 to any t_j in the interval $t_0 \le t \le b$ satisfies

$$\text{Error} \le Ch,$$

where C is a constant that depends on $f(t, y)$, its partial derivatives, the initial condition, and the interval, but not on h. Moreover, it can be shown that the error is actually proportional to h. Because of this, the Euler Method is called a *first order method.* Note that in the example, cutting the step size in half had the effect of cutting the error approximately in half, as expected for a first order method.

It is also useful to know the *local error.* Suppose $u(t)$ is the solution of the differential equation satisfying $u(t_j) = y_j$. Then the local error in stepping from t_j to t_{j+1} is defined to be

$$e_{j+1} = u(t_{j+1}) - y_{j+1};$$

i.e., the error made in one step assuming the solution value at t_j is y_j. Using Taylor's formula it is easily shown that, for the Euler Method,

$$e_{j+1} = \frac{1}{2}u''(\bar{t}_j)h^2,$$

for some $\bar{t}_j \in (t_j, t_{j+1})$. Thus the local error is proportional to h^2. Because local error provides a simple comparison of methods and is used in the design of numerical solution software, we will state the local error for each method we discuss. The error discussed in the previous paragraph is called *global error*, in order to distinguish it from local error.

The relation between local and global error can be understood intuitively in the following way. We begin by noting that

$$\text{Global Error at } t_{j+1} = y(t_{j+1}) - y_{j+1} = (y(t_{j+1}) - u(t_{j+1})) + (u(t_{j+1}) - y_{j+1}).$$

The second term on the right-hand side of this equation is the local error, which is proportional to h^2. The first term is the difference at t_{j+1} of the two solutions of the differential equation with values $y(t_j)$ and y_j at t_j. The size of this term depends on the size of $y(t_j) - y_j$ and on the differential equation, specifically on whether the difference between $y(t_j)$ and y_j (which serve as initial values at t_j) leads to a smaller or larger difference in the values of the solutions at t_{j+1}, *i.e.*, on whether the differential equation is stable or unstable. These terms were defined in Chapter 5.

Suppose the differential equation is stable. Then its solutions are fairly insensitive to initial values, so that $y(t_{j+1}) - y_{j+1}$ is about the same size as or smaller than $y(t_j) - y_j$. In this situation we see that the accumulated, global, error is approximately the sum of the local errors. To be specific, in stepping from t_0 to t_j, we make j local errors, each of which is proportional to h^2. Since $j \leq n$, we thus have an accumulated error that is no greater than a constant times

$$nh^2 = \frac{b - t_0}{h}h^2 = (b - t_0)h;$$

i.e., the global error is bounded by a constant times h.

Next, suppose the differential equation is unstable, so that $u(t_{j+1}) - y_{j+1}$ may be larger than $y(t_j) - y_j$. Then the global error at t_{j+1} is the sum of this larger quantity and the local error at t_{j+1}. Thus, in this situation, the global error depends on the stability of the differential equation as well as on the numerical method, and cannot be completely assessed by the simple calculation sketched above. It can still be proved that, as stated earlier, the global error is proportional to h. In this result stability influences the constant of proportionality: if the differential equation is stable, the constant is of moderate size; if unstable, the constant may be large. The effect of stability and instability on numerical solutions will be discussed further in Section 7.4.

The Euler Method in MATLAB. The Euler Method can be implemented using the following function M-file, named `myeuler.m`:

```
function [t, y] = myeuler(f, tinit, yinit, b, n)

% Euler approximation for initial value problem,
% dy/dt = f(t, y), y(tinit) = yinit.
% Approximations y(1),..., y(n + 1) are calculated at
% the n + 1 points t(1),..., t(n + 1) in the interval
% [tinit, b].  The right-hand side of the differential
% equation is defined as an anonymous function f.

% Calculation of h from tinit, b, and n.

h = (b -  tinit)/n;

% Initialize t and y as length n + 1 column vectors.
```

```
t = zeros(n + 1, 1);
y = zeros(n + 1, 1);

% Calculation of points t(i) and the corresponding
% approximate values y(i) from the Euler Method formula.

t(1) = tinit;
y(1) = yinit;
for i = 1:n
    t(i + 1) = t(i) + h;
    y(i + 1) = y(i) + h*f(t(i), y(i));
end
```

To run this M-file, define an anonymous function **f** corresponding to the right-hand side of the differential equation, choose appropriate values for the last four arguments, and type **[t, y] = myeuler(f, t0, y0, b, n)**. If the right-hand side is defined via the M-file, `func.m`, type **[t, y] = myeuler(@func, t0, y0, b, n)**. You can use the M-file **myeuler** as a template for implementing other fixed step size methods, two of which are discussed next.

The vector **t** will contain the points $t_0, t_1, \ldots, t_n$, and the vector **y** will contain the corresponding approximations $y_0, y_1, \ldots, y_n$. These can be displayed in tabular form with the command **[t y]**. Note that **t(i)** corresponds to t_{i-1}, and **y(i)** to y_{i-1}. This correspondence is necessary because the first element in a MATLAB vector is indexed by **1**, and not by **0**; in particular, the first element in **t** is **t(1)**, and the first in **y** is **y(1)**.

The initialization of **t** and **y** as column vectors is not essential, but it serves two purposes. First, **[t y]** is in tabular form because **t** and **y** are columns. Second, initializing **t** and **y** causes the program to run more efficiently.

To approximate the solution at a point t that is not a t_i, we would have to *interpolate*. A simple way to do this is to "connect the dots" by drawing straight line segments from each point (t_i, y_i) to the next point (t_{i+1}, y_{i+1}); thus $y_a(t)$ would be the piecewise linear function connecting the computed points (t_i, y_i). (A more sophisticated way to interpolate is described in Section 8.6.1.) To graph $y_a(t)$, type **plot(t, y)**.

7.2.2 The Improved Euler Method

We again start with the tangent line approximation, replacing the slope $y'(t_i)$ by the average of the two slopes $y'(t_i)$ and $y'(t_{i+1})$. This yields

$$\begin{aligned} y(t_{i+1}) &\approx y(t_i) + h\,\frac{y'(t_i) + y'(t_{i+1})}{2} \\ &= y(t_i) + h\,\frac{f(t_i, y(t_i)) + f(t_{i+1}, y(t_{i+1}))}{2}. \end{aligned} \tag{7.3}$$

We now use the approximation $y(t_i) \approx y_i$ and the previous Euler approximation $y(t_{i+1}) \approx y_i + hf(t_i, y_i)$ to get

$$\begin{aligned} y(t_{i+1}) &\approx y_i + h\,\frac{f(t_i, y_i) + f(t_{i+1}, y_i + hf(t_i, y_i))}{2} \\ &= y_i + h\,\frac{y_i' + f(t_{i+1}, y_i + hy_i')}{2} \\ &= y_{i+1}, \end{aligned}$$

where $y_i{}' = f(t_i, y_i)$. Given the initial value y_0, this analysis leads to the recursive formula

$$y_{i+1} = y_i + h\,\frac{y_i' + f(t_{i+1}, y_i + hy_i')}{2}, \qquad i = 0, 1, \ldots, n-1.$$

Example 7.2 We again consider problem (7.1) and find an approximation to $y(0.3)$ with $h = 0.1$. We obtain

$$\begin{aligned} y_1 &= y_0 + h\,\frac{y_0' + f(t_1, y_0 + hy_0')}{2} \\ &= y_0 + \frac{h}{2}\left(\frac{t_0}{y_0} + \frac{t_1}{y_0 + ht_0/y_0}\right) \\ &= 1.005 \\ y_2 &= 1.019828 \\ y_3 &= 1.044064. \end{aligned}$$

So Error $= |y(0.3) - y_3| = 0.000033\ldots$. Note that the error in the Improved Euler Method with $h = 0.1$ is considerably less than the error in the Euler Method with $h = 0.05$.

The local error for the Improved Euler Method is proportional to h^3.

7.2.3 The Runge-Kutta Method

Given the initial value y_0, the fourth order Runge-Kutta Method is defined recursively by

$$y_{i+1} = y_i + \frac{h}{6}(k_1 + 2k_2 + 2k_3 + k_4),$$

where

$$\begin{aligned} k_1 &= f(t_i, y_i) \\ k_2 &= f(t_i + \frac{h}{2}, y_i + \frac{h}{2}k_1) \\ k_3 &= f(t_i + \frac{h}{2}, y_i + \frac{h}{2}k_2) \\ k_4 &= f(t_{i+1}, y_i + hk_3). \end{aligned}$$

Note that a weighted average of the "slopes" k_1, k_2, k_3, k_4 is used in the formula for y_{i+1}.

Example 7.3 Consider problem (7.1) and again find an approximation to $y(0.3)$, but now let $h = 0.3$ and take just one step. Then

$$\begin{aligned} k_1 &= f(0,1) = 0 \\ k_2 &= f(0+0.15, 1+0.15(0)) = 0.15 \\ k_3 &= f(0.15, 1+0.15(0.15)) = 0.146699 \\ k_4 &= f(0.3, 1+0.3(0.146699)) = 0.287354 \\ y_1 &= 1 + \frac{0.3}{6}(0 + 2(0.15 + 0.146699) + 0.287354) \\ &= 1.044038 \approx y(0.3). \end{aligned}$$

So Error $= |y(0.3) - y_1| = 0.000007\ldots$. We see that the error in the Runge-Kutta Method with $h = 0.3$ is much less than the error in the Euler Method, or even in the Improved Euler Method, with $h = 0.05$. The local error for the Runge-Kutta Method is proportional to h^5.

Remark 7.1 The methods we have discussed can be applied to any first order equation. Furthermore, they can be applied using different step sizes at each step.

7.2.4 Inside `ode45`

Software for the numerical solution of ordinary differential equations can be based on the methods we have presented, as well as on other methods. The numerical solver **`ode45`** combines a fourth order method and a fifth order method, both of which are similar to the fourth order Runge-Kutta Method discussed above. It varies the step size, choosing the step size at each step in an attempt to achieve the desired accuracy. All efficient modern numerical ODE solvers use variable step size methods. The solver **`ode45`** is suitable for a wide variety of initial value problems.

The solver **`ode45`** produces approximate solution values at the sequence of points determined by its selection of step sizes; we refer to these points as internal time points. Suppose you have requested approximate solution values at certain specific points, as, for example, with **`ode45(f, 0:0.2:2, 1)`**. The points **`0:0.2:2`** will usually not coincide with the internal time points, and interpolation must be used to get approximate solution values at the points **`0:0.2:2`**. The interpolation procedure connects the approximate solution values at the internal time points, not with straight lines (linear polynomials), as mentioned for the Euler Method, but with higher degree polynomials. The solver **`ode45`** attempts to perform both processes—the calculation of the approximate solution values at the internal time points and the interpolation process—in such a way that the total error in both processes does not exceed the desired level, by default 10^{-3}. Finally, when **`ode45`** is used without points being specified, as, for example, with **`ode45(f, [0 2], 1)`**, approximate solution values are computed at the internal time points, and then by interpolation at three equally spaced points between each pair of these points. In the graph produced by this command, the approximate solution values at all of these points are indicated by circles, and the centers of the circles are then connected with line segments (see Figure 7.1).

MATLAB includes two other numerical solvers, **ode23** and **ode113**, that can be used with the problems in this book. For some initial value problems (referred to as *stiff* problems), the solvers **ode45**, **ode23**, and **ode113** are less satisfactory than methods that have been designed specifically to solve stiff problems. MATLAB includes four such solvers, **ode15s**, **ode23s**, **ode23**, and **ode23bt**. Additional information on all of these solvers can be found in the online help and in the browser. For additional information on numerical methods and their software implementations, we refer to D. Kahaner, C. Moler, and S. Nash, **Numerical Methods and Software**, Prentice Hall, Inc., 1989. Especially recommended are two books that focus on numerical methods in MATLAB: L. Shampine, I. Gladwell, and S. Thompson, **Solving ODEs with MATLAB**, Cambridge University Press, 2003, and C. Moler, **Numerical Computing with MATLAB**, SIAM, 2004, also available at: `http://www.mathworks.com/moler/`.

7.2.5 Round-off Error

The type of error discussed above is called *discretization error*, and would be present even if one could retain an infinite number of digits. In addition, there is *round-off error*, which arises because the computer uses a fixed, finite number of digits. Letting $\tilde{y}_a(x)$ denote the actual computed approximate solution, which includes round-off error, the total error can be written

$$\begin{aligned} y(t) - \tilde{y}_a(t) &= (y(t) - y_a(t)) + (y_a(t) - \tilde{y}_a(t)) \\ &= \text{Discretization Error + Round-off Error.} \end{aligned}$$

Since most computers that run MATLAB carry 16 digits, the major portion of the error will be the discretization error. Thus, for most problems, round-off error can be safely ignored. For this reason we will not distinguish between $y_a(t)$ and $\tilde{y}_a(t)$.

7.3 Controlling the Error in ode45

As indicated above, **ode45** attempts to provide an approximate solution with error approximately $\leq 10^{-3}$. A more accurate approximate solution can be obtained by using an optional fourth argument to **ode45**. This argument is created using the **odeset** command. For example, to compute an approximate solution to the initial value problem (7.1) at $t = 2$, requesting 12-digit accuracy, type

```
>> options = odeset('AbsTol', 1e-12, 'RelTol', 1e-12);
>> [t, ya] = ode45(f, [0 2], 1, options);
>> format long
>> ya(end)
```

These commands produce the output 2.23606797749981, which approximates the exact value $\sqrt{5}$ to 12 digits.

AbsTol specifies absolute error, and **RelTol** specifies relative error. With the settings **1e-**a for **AbsTol** and **1e-**r for **RelTol**, where a and r are positive integers, **ode45** tries to approximate $y(t)$ with error less than $\max(10^{-a}, |y(t)|10^{-r})$. Without these options, MATLAB uses the default settings **1e-6** for **AbsTol** and **1e-3** for **RelTol**. The default values will be satisfactory for most of the problems in this book. For other problems, however, greater accuracy will be required.

7.4 Reliability of Numerical Methods

We have claimed that **ode45**, when used with the default options, leads to approximately 3-digit accuracy. More precisely, the default options ensure that the *local error*, *i.e.*, the error made in a step of length h, is at worst approximately 10^{-3}. One is generally more interested in the global error, defined above to be the cumulative error committed in taking the necessary number of steps to get from the initial point t_0 to b. This error cannot be controlled completely by the numerical method because it depends on the differential equation as well as on the numerical method. The key issue is whether the differential equation is stable or unstable (see Chapter 5). In the earlier discussion of local and global error for the Euler Method, we saw that the global error grows only moderately if the differential equation is stable, but grows much faster if the equation is unstable. We will now illustrate the effect of stability and instability on accuracy by considering an example for which **ode45** gives poor results.

Example 7.4 Consider the initial value problem

$$\frac{dy}{dt} = y - 3e^{-2t}, \qquad y(0) = 1. \tag{7.4}$$

The exact solution is $y(t) = e^{-2t}$.

Now we find the numerical solution using **ode45**, and plot it (with a dashed line) together with the exact solution (with a solid line):

```
>> f = @(t, y) y - 3*exp(-2*t);
>> [t, ya1] = ode45(f, [0 12], 1);
>> plot(t, ya1, '--')
>> hold on
>> ezplot('exp(-2*t)', [0 12])
>> hold off
>> axis([0 12 -1 1])
>> xlabel 't'
>> ylabel 'y'
>> title ''
```

We see from Figure 7.3 that the graphs of **ya1** and y are indistinguishable from 0 to about 6, but then the curves separate sharply. Their values at $t = 12$, for instance, are very different (**ya1(end)** $= -1.29\ldots$ and $y(12) = e^{-24} = 3.77\cdots\times 10^{-11}$).

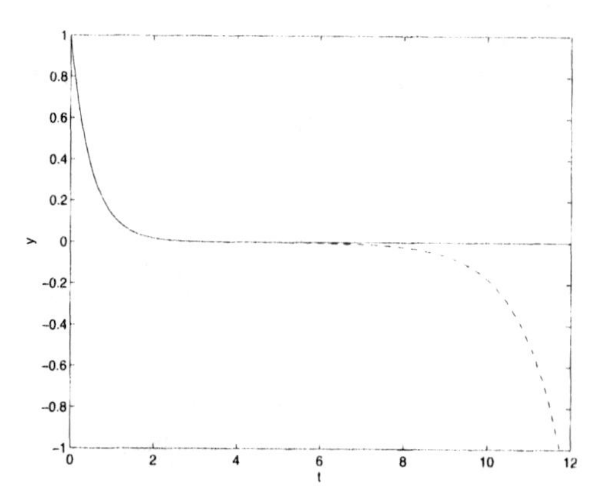

Figure 7.3: Exact (solid line) and Approximate (dashed line) Solutions

How can this failure be explained? The solution to our differential equation with initial condition $y(0) = 1 + \epsilon$ is $y(t) = e^{-2t} + \epsilon e^t$. Figure 7.4 shows a plot of the solutions corresponding to initial values $1-(0.5\times 10^{-5}), 1-(0.25\times 10^{-5}), 1, 1+(0.25\times 10^{-5}), 1+(0.5 \times 10^{-5})$.

We see that as t increases, the solutions with initial values slightly different from 1 separate very sharply from the solution with initial value exactly 1; those with initial values greater than 1 approach $+\infty$, and those with initial values less than 1 approach $-\infty$. Now **ode45**, like any numerical method, introduces errors. These errors—whether discretization errors or round-off errors—have caused the numerical solution to jump to a solution that started just below 1. Once on such a solution, the numerical solution will follow it or another such solution. Since these solutions approach $-\infty$, the numerical solution does likewise.

Could we have anticipated this failure? Recall the discussion of stable and unstable differential equations in Chapter 5. The solutions of unstable equations are very sensitive to their initial values, and hence a numerical method will have trouble following the solution it is attempting to calculate. Recall also that an equation $dy/dt = f(t,y)$ is unstable if $\partial f/\partial y > 0$, and stable if $\partial f/\partial y \leq 0$. For our example, $f(t,y) = y - 3e^{-2t}$ and $\partial f/\partial y = 1 > 0$. So we are trying to approximate the solution of an unstable differential equation, and we should not be surprised that we have trouble, especially over long intervals. (By contrast, note that for the initial value problem (7.1) above, $\partial f/\partial y = -t/y^2 < 0$ for $t > 0$.) All this suggests caution when dealing with unstable equations. Caution is always recommended when modeling physical problems with unstable equations, especially over long intervals (*cf.* the discussion in Section 5.3).

In assessing the reliability of a numerically computed solution there is another test one can make. Suppose we have made a calculation using the default options, as we have done with our example. Then one can do another calculation, using a more stringent accuracy requirement, and compare the results. If they are nearly the same, one can have reasonable

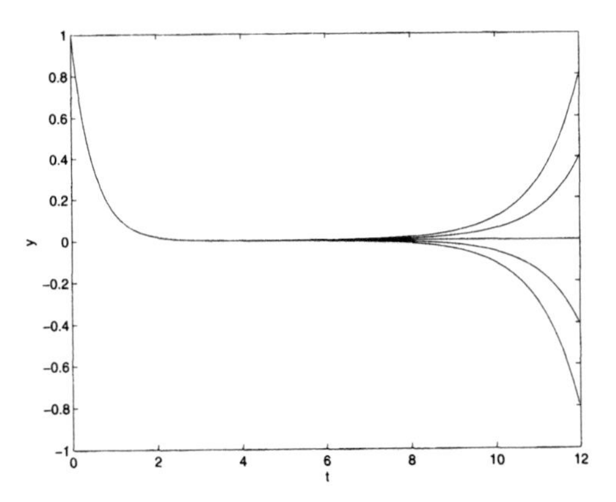

Figure 7.4: Exact Solution with Five Different Initial Values

confidence that they are both accurate; but if they differ substantially, then one should suspect that the first calculation is not accurate.

Let's do this with our example. Specifically, we calculate and plot the numerical solution, **ya2**, satisfying more stringent accuracy requirements, together with the first numerical solution, **ya1**, and the exact solution, y. We display the exact solution with a solid line, **ya1** with a dotted line, and **ya2** with a dashed line.

```
>> f = @(t, y) y - 3*exp(-2*t);
>> [t, ya1] = ode45(f, [0 17], 1);
>> plot(t, ya1, ':')
>> hold on
>> options = odeset('AbsTol', 1e-8, 'RelTol', 1e-5);
>> [t, ya2] = ode45(f, [0 17], 1, options);
>> plot(t, ya2, '--')
>> ezplot('exp(-2*t)', [0 17])
>> hold off
>> axis([0 17 -1 1])
>> xlabel 't'
>> ylabel 'y'
>> title ''
```

We see from Figure 7.5 that the second numerical solution is accurate over a longer interval, out to about 13, whereas the first was accurate out to about 6. In particular, the two numerical solutions differ after about 6. Even if we didn't know the exact solution, we could have concluded that the numerical solution is accurate out to, but not beyond, approximately 6.

In summary, we have learned that local accuracy—the accuracy that a numerical method can control—leads to reasonable global accuracy if the differential equation is stable, and

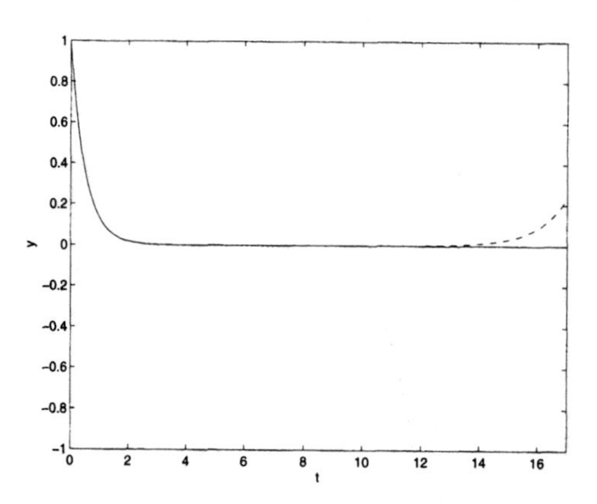

Figure 7.5: Effect of Increased Accuracy

that the reliability of a numerical solution can also be checked by computing another numerical solution with a more stringent accuracy requirement.

Several additional ways to use **ode45** are discussed in Section 8.6.

Chapter 8

Features of MATLAB

This chapter is a continuation of Chapters 3 and 4. We describe some of the more advanced features of MATLAB, focusing on MATLAB commands and techniques that are most useful for studying differential equations.

8.1 Data Classes

Every variable you define in MATLAB, as well as every input and output to a command, is an *array* of data belonging to a particular *class*. In this book we use primarily four types of data: floating point numbers, symbolic expressions, character strings, and function handles. We introduced each of these types in Chapter 3. In Table 8.1, we list for each type of data its class (as given by the command **whos**) and how you can create it.

Type of data	Class	Created by
Floating point	`double`	typing a number
Symbolic	`sym`	using **sym** or **syms**
Character string	`char`	typing a string inside single quotes
Function handle	`function_handle`	using **@**

Table 8.1: Four MATLAB Data Classes

You can think of an array as a two-dimensional grid of data; MATLAB allows arrays to have more than two dimensions, but we will not use this feature. A single number (or symbolic expression, or function handle) is regarded by MATLAB as a 1-by-1 array, sometimes called a *scalar*. A 1-by-n array is called a *row vector*, and an m-by-1 array is called a *column vector*. (A string is actually a row vector of characters.) An m-by-n array of numbers is called a *matrix*; see Section 8.4 below. You can see the class and array size of every variable you have defined by typing **whos** (see Section 3.6). The set of variable definitions shown by **whos** is called your *Workspace*.

In order to use MATLAB commands effectively, you must pay close attention to the class of data each command accepts as input and returns as output. The input to a command consists of one or more arguments separated by commas; often, some arguments are optional. When the input arguments are all strings, you can omit all punctuation; then MATLAB interprets the first word as the name of the command and the remaining words as its string arguments. Thus, **hold on** is equivalent to **hold('on')**. Some commands, like **whos**, do not require any input. The online help (see Section 2.5) for each command usually tells what classes of inputs the command expects, as well as what class of output it returns.

Many commands allow more than one class of input, though sometimes only one data class is mentioned in the online help. This flexibility can be a convenience in some cases and a pitfall in others. For example, as we saw in Section 3.8, **ezplot** allows you to specify the function to be plotted as a string, a symbolic expression, or a function handle. On the other hand, suppose that you have previously defined **x = [5 7]**, and now you attempt to take the derivative of $\log x$ by typing

```
>> diff(log(x))
ans =
    0.3365
```

The reason you don't get an error message is that **diff** is the name of a function that computes numerical difference quotients as well as symbolic derivatives. Since **x** is a vector, so is **log(x)**, and the numerical version of **diff** is applied. In general, when the output of a command doesn't look right to you, and MATLAB gives no error or warning message, you should consider the data class of the input arguments you supplied, and check the online help for the command if you are not sure what data classes are appropriate.

Remark 8.1 Typing **help diff** only shows you the help text for the numerical version of the command, but gives a cross-reference to the symbolic version at the bottom. If you want to see the help text for the symbolic version instead, type **help sym/diff**.

Sometimes you need to convert one data class into another in order to prepare the output of one command to serve as the input for another. For example, we have used **double** to convert symbolic expressions to floating-point numbers and **sym** to convert numbers or strings to symbolic expressions. The commands **num2str** and **str2num** convert between numbers and strings, while **char** converts a symbolic expression to a string. You can also use **vectorize** to convert a symbolic expression to a vectorized string; it adds a **.** before every *, /, and ^ in the expression.

8.1.1 String Manipulation

Often it is useful to concatenate two or more strings together. The simplest way to do this is to use MATLAB's vector notation, keeping in mind that a string is a "row vector" of characters. For example, typing **[string1 string2]** combines **string1** and **string2** into one string.

Here is a useful application of string concatenation. You may need to define a string variable containing an expression that takes more than one line to type. (In most circumstances you can continue your input onto the next line by typing **...** followed by ENTER or RETURN, but this is not allowed in the middle of a string.) Break the expression into smaller parts, assign a variable to each part, and concatenate the variables, as in:

```
>> lhs = 'left hand side of equation';
>> rhs = 'right hand side of equation';
>> eqn = [lhs ' = ' rhs]
eqn =
left hand side of equation = right hand side of equation
```

8.1.2 Symbolic and Floating Point Numbers

We mentioned above that you can convert between symbolic numbers and floating point numbers with **double** and **sym**. Numbers that you type are, by default, floating point. However, if you mix symbolic and floating point numbers in an arithmetic expression, the floating point numbers are automatically converted to symbolic. This explains why you can type **syms x** and then **x^2** without having to convert **2** to a symbolic number. Here is another example:

```
>> a = 2
a =
   2
>> b = a/sym(3)
b =
2/3
```

MATLAB was designed so that some floating point numbers are restored to their exact values when converted to symbolic. Integers, rational numbers with small numerators and denominators, square roots of small integers, the number π, and certain combinations of these numbers are so restored. For example:

```
>> c = sqrt(5)
c =
   2.2361
>> sym(c)
ans =
sqrt(5)
```

Since it is difficult to predict when MATLAB will preserve exact values, it is best to suppress the floating point evaluation of a numeric argument to **sym** by enclosing it in single quotes to make it a string, *e.g.*, **sym('1 + sqrt(5)')**. We will see below another way in which single quotes suppress evaluation.

8.1.3 Structures

Another data class that you will occasionally encounter in MATLAB is a *structure*. An example occurred in Section 3.7:

```
>> sol = solve('x + y^2 = 2', 'y - 3*x = 7')
sol =
   x: [2x1 sym]
   y: [2x1 sym]
```

If you type **whos sol**, you will see that the data class of **sol** is "struct array". A structure can contain any number of named variables, called "fields", together with their values. In this case, **sol** has two fields, **x** and **y**, and each is a vector containing two symbolic numbers. To access a particular field, type the name of the structure followed by the name of the field, with a period in between. Thus, **sol.x** is a vector containing the x values of the two solutions of the system of equations given to **solve**, and **sol.x(1)** is the first number in this vector.

More generally, a structure can contain multiple data classes. For example, one field can be a string and another can be an array of floating point numbers, or a function handle. A field can even be another structure. An example of a more complex structure like this is given by **ode45** when its output is assigned to a single variable; see Section 8.6.1 below. Although in this book we will use structures only when they occur as MATLAB output, you can define your own structures with the command **struct**; see the online help for examples.

8.2 Functions and Expressions

We have used the terms *expression* and *function* without carefully discussing the distinction between the two. Strictly speaking, if we define $f(x) = x^2 + 1$, then f (written without any particular input) is a function while $f(x)$ and $x^2 + 1$ are expressions involving the variable x. In mathematical discourse, we often blur this distinction by calling $f(x)$ or $x^2 + 1$ a function, but in MATLAB the difference between functions and expressions is important.

In MATLAB, an expression can belong to either the string or symbolic class of data. Consider the following example.

```
>> f = 'x^2 + 1'; f(3)
ans =
2
```

This result may be puzzling if you are expecting **f** to act like a function. Since **f** is a string, **f(3)** denotes the third character in **f**, which is **2**. (Notice that unlike numeric output, this output is not indented from the left margin; it is a string consisting of one character.) Typing **f(5)** would yield **+** and **f(-1)** would produce an error message.

One way to get the desired answer in the example above is with the command **eval**, which evaluates a string or symbolic expression in light of whatever values you have assigned to variables:

```
>> x = 3; eval(f)
ans =
    10
```

Typing **f** rather than **eval(f)** above would have simply output the expression **f** without substituting **3** for **x** and evaluating the result. For a similar reason, we used **eval** in Chapter 5 to define a function from the output of a symbolic command. You can also substitute a value into an expression using **subs**, as described in Section 3.3.1.

You have learned two ways to define your own functions: on the command line, using **@** (to create an "anonymous function") or **inline** (see Section 3.5); and by creating a function M-file (see Section 4.2.2). Anonymous and inline functions are most useful for defining simple functions that can be expressed in one line and for turning the output of a symbolic command into a function. Function M-files are useful for defining functions that require several intermediate commands to compute the output. Most MATLAB commands are actually M-files, and you can peruse them for ideas to use in your own M-files—to see the M-file for, say, the command **mean** you can enter **type mean**. See also Section 8.3 below.

Some commands, such as **ode45**, require their first argument to be a function. If you type an expression instead, you will get an error message. On the other hand, most symbolic commands, such as **solve** or **dsolve**, require their first argument to be either a string or a symbolic expression, and not a function.

An important difference between strings and symbolic expressions is that MATLAB automatically substitutes user-defined functions and variables into symbolic expression that you type, but not into strings. (This is another sense in which the single quotes you type around a string suppress evaluation.) Consider the following example:

```
>> h = @(t) t.^3; int('h(t)', 't')
Warning: Explicit integral could not be found.
> In sym.int at 58
  In char.int at 9

ans =
int(h(t),t)
```

MATLAB cannot evaluate the integral because within a string, **h** is regarded as an unknown function. If you make the input to **int** symbolic, then MATLAB substitutes the previous definition of **h** before performing the integration:

```
>> syms t; int(h(t), t)
ans =
1/4*t^4
```

8.3 More about M-files

Files containing MATLAB statements are called M-files. There are two kinds of M-files: *function M-files*, which accept arguments and produce output, and *script M-files*, which execute a series of MATLAB statements. Earlier we created and used both types. In this section we present additional information on M-files.

8.3.1 Variables in Script M-files

When you run a script M-file, the variables you use and define belong to your Workspace; *i.e.*, they take on any values you assigned earlier in your MATLAB session, and they persist after the script finishes executing. Consider the following script M-file, called `scriptex1.m`:

```
u = [1 2 3 4];
```

Typing **`scriptex1`** assigns the given vector to **`u`** but displays no output. Now consider another script, called `scriptex2.m`:

```
length(u)
```

If you have not previously defined **`u`**, then typing **`scriptex2`** will produce an error message. However, if you type **`scriptex2`** after running **`scriptex1`**, then the definition of **`u`** from the first script will be used in the second script and the output **`ans = 4`** will be displayed.

If you don't want the output of a script M-file to depend on any earlier computations in your MATLAB session, put the line **`clear all`** near the beginning of the M-file, as we suggested in Section 4.2.1.

Sometimes you might want your script M-file to accept inputs interactively. You can arrange this with the command **`input`**. For example, if you include the line **`var = input('Input var here:  ')`** in your script, when MATLAB gets to that point it will print "`Input var here:  `" and pause while you type the value to be assigned to **`var`**.

You can also format your output in a more descriptive manner than **`ans = 4`** by using **`disp`**:

```
>> disp('The length of u is:'), disp(length(u))
The length of u is:
     4
```

To get the output all on one line, you can convert the numeric output to a string with **`num2str`** and concatenate the strings:

```
>> disp(['The length of u is ' num2str(length(u)) '.'])
The length of u is 4.
```

8.3.2 Variables in Function M-files

The variables used in a function M-file are *local*, meaning that they are unaffected by, and have no effect on, the variables in your Workspace. Consider the following function M-file, called `sq.m`:

```
function z = sq(x)
% sq(x) returns the square of x
z = x.^2;
```

Typing **`sq(3)`** produces the answer **`9`**, whether or not **`x`** or **`z`** is already defined in your Workspace, and does not change their definitions.

8.3.3 Structure of Function M-files

The first line in a function M-file is called the *function definition line*; it defines the function name, as well as the number and order of input and output arguments. Following the function definition line, there can be several comment lines that begin with a percent sign (**%**). These lines are called *help text* and are displayed in response to the **help** command. In the M-file `sq.m` above, there is only one line of help text; it is displayed when you type **help sq**. The remaining lines constitute the *function body*; they contain the MATLAB statements that calculate the function values. In addition, there can be comment lines (lines beginning with **%**) anywhere in an M-file. All statements in a function M-file that normally produce output should end with a semicolon to suppress the output.

Function M-files can have multiple input and output arguments. Here is an example, called `polarconvert.m`, with two input and two output arguments.

```
function [r, theta] = polarconvert(x, y)
% polarconvert(x, y) returns the polar coordinates
% of the point with rectangular coordinates (x, y)
r = sqrt(x.^2 + y.^2);
theta = atan2(y, x);
```

If you type **polarconvert(3, 4)**, only the first output argument, **r**, is returned and stored in **ans**; in this case, the answer is **5**. To see both outputs, you must assign them to variables enclosed in square brackets:

```
>> [r, theta] = polarconvert(3, 4)
r =
    5


theta =
    0.9273
```

If you assign the output to only one variable, it will receive the first output argument, regardless of the variable's name; typing **theta = polarconvert(3, 4)** will assign **5** to **theta**. (Remember, the names of variables inside a function M-file do not affect variables with the same name in your workspace.)

8.4 Matrices

A *matrix* is a rectangular array of numbers. Row and column vectors, which we discussed in Section 3.4, are examples of matrices. Consider the 3×4 matrix

$$A = \begin{pmatrix} 1 & 2 & 3 & 4 \\ 5 & 6 & 7 & 8 \\ 9 & 10 & 11 & 12 \end{pmatrix}.$$

It can be entered in MATLAB with the command

```
>> A = [1,2,3,4; 5,6,7,8; 9,10,11,12]
```

```
A =
     1     2     3     4
     5     6     7     8
     9    10    11    12
```

Note that the matrix *elements* in any row are separated by commas, and the rows are separated by semicolons. The elements in a row can also be separated by spaces. The dimensions of a matrix can be found with **size**:

```
>> size(A)
ans =
   3     4
```

If two matrices **A** and **B** are the same size, their sum is obtained by typing **A + B**. You can also add a scalar (a single number) to a matrix; **A + c** adds **c** to each element in **A**. Likewise, **A - B** represents the difference of **A** and **B**, and **A - c** subtracts the number **c** from each element of **A**. If **A** and **B** are multiplicatively compatible; *i.e.*, if **A** is $n \times m$ and **B** is $m \times \ell$, then their product **A*B** is $n \times \ell$. The product of a number **c** and the matrix **A** is given by **c*A**, and **A'** represents the conjugate transpose of **A**. For more information, see the online help for **ctranspose** and **transpose**.

If **A** and **B** are the same size, then **A.*B** is the *element-by-element* product of **A** and **B**, *i.e.*, the matrix whose i, j element is the product of the i, j elements of **A** and **B**. Likewise, **A./B** is the element-by-element quotient of **A** and **B**, and **A.^c** is the matrix formed by raising each of the elements of **A** to the power **c**. More generally, if **f** is one of the built-in mathematical functions in MATLAB, or is a user-defined function that is vectorized, then **f(A)** is the matrix obtained by applying **f** element-by-element to **A**. See what happens when you type **sqrt(A)**, where **A** is the matrix defined above. In Chapter 3 we used such element-by-element operations in graphing.

Recall that **x(3)** is the third element of a vector **x**. Likewise, **A(2, 3)** represents the 2, 3 element of **A**, *i.e.*, the element in the second row and third column. You can specify submatrices in a similar way. Typing **A(2, [2 4])** yields the second and fourth elements of the second row of **A**. To select the second, third, and fourth elements of this row, type **A(2, 2:4)**. The submatrix consisting of the elements in rows 2 and 3 and in columns 2, 3, and 4 is generated by **A(2:3, 2:4)**. A colon by itself denotes an entire row or column. For example, **A(:, 2)** denotes the second column of **A**, and **A(3, :)** yields the third row of **A**.

MATLAB has several commands that generate special matrices. The commands **zeros(n, m)** and **ones(n, m)** produce $n \times m$ matrices of zeros and ones, respectively. Also, **eye(n)** represents the $n \times n$ identity matrix. In Section 7.2.1, we used **zeros** in the M-file myeuler.m to initialize the output vectors.

8.4.1 Solving Linear Systems

Suppose **A** is a nonsingular $n \times n$ matrix and **b** is a column vector of length n. Then typing **x = A\b** numerically computes the unique solution to **A*x = b**. Type **help mldivide** for more information.

If either **A** or **b** is symbolic rather than numeric, then `x = A\b` computes the solution to `A*x = b` symbolically. To calculate a symbolic solution when both inputs are numeric, type `x = sym(A)\b`.

8.4.2 Calculating Eigenvalues and Eigenvectors

The eigenvalues of a square matrix **A** are calculated with `eig(A)`. If you specify two output variables, as in `[U, R] = eig(A)`, then MATLAB calculates both the eigenvalues and eigenvectors. The eigenvalues are the diagonal elements of the diagonal matrix **R**, and the columns of **U** are the eigenvectors. Here are two examples illustrating the use of `eig`.

```
>>  A = [3 -2 0; 2 -2 0; 0 1 1]; eig(A)
ans =
      1
     -1
      2
>>  [U, R] = eig(A)
U =
           0   -0.4082   -0.8165
           0   -0.8165   -0.4082
      1.0000    0.4082   -0.4082

R =
      1     0     0
      0    -1     0
      0     0     2
```

The eigenvector in the first column of **U** corresponds to the eigenvalue in the first column of **R**, and so on. These are numerical values for the eigenpairs. To get symbolically calculated eigenpairs, type `[U, R] = eig(sym(A))`.

8.5 Graphics

In Section 3.8 we discussed the plotting commands `ezplot`, `plot`, and `contour`, as well as the basic commands for adjusting and labeling the axes of a graph. These commands, along with `quiver`, which we introduced in Chapter 6 to plot direction fields, are the main graphics commands you will need for this course. In this section we describe additional commands for manipulating graphics. In many cases, we describe how to do this both from the MATLAB prompt and with the mouse. While you may find the mouse operations simpler, the text commands allow you to modify graphics from an M-file.

8.5.1 Figure Windows

When you execute the first plotting command in a given MATLAB session, the graph appears in a new window labeled "Figure 1". Subsequent graphics commands either modify or replace the graph in this window. You have seen that **hold on** directs that new plotting commands should add to, rather than replace, the current graph. If instead you want to create a new graph in a separate window while keeping the old graph on your screen, type **figure** to open a new window, labeled "Figure 2". Alternatively, you can select a New Figure from the **File** menu in either your main MATLAB window or the first figure window. Subsequent graphics commands will affect only this window, until you change your *current figure* again with another **figure** command, or by bringing another figure to the foreground with the mouse. Typing **figure** alone will create a third figure window, while typing **figure(1)** will switch the current window back to "Figure 1". You can delete figure windows with the command **close**; see the online help for details.

Each figure window has a tool bar underneath its menu bar with shortcuts for several menu items, including on the left the usual icons for opening, saving, and printing figures. Near the middle, there are two icons with plus and minus signs, for zooming in and out. Click on the icon with a plus sign, and then click on a point in your graph to zoom in near that point. You can click and zoom multiple times; each zoom changes the scale on both axes by a factor of roughly 2.5. However, don't click too fast, because double-clicking resets the graph to its original state. Clicking on the icon with the minus sign allows you to zoom out gradually, and clicking on the icon of a hand allows you to click and drag the graph to pan both horizontally and vertically within the current axes. More zooming options are available by right-clicking in the figure window while zooming, by using the **Options** submenu of the **Tools** menu, or by using the command **zoom** (see its online help).

Clicking the next icon to the right of the hand icon allows you to rotate 3D graphics. More 3D manipulations are available by right-clicking in the figure window, and by selecting the Camera Toolbar from the **View** menu.

Clicking the next icon to the right of the rotate icon enables the Data Cursor, which allows you to display the coordinates of a point on a curve by clicking on or near the curve. Right-clicking in the figure window gives several options; changing Selection Style to Mouse Position will allow you to click on an arbitrary point on the curve rather than just a data point. (Remember that the curves plotted by MATLAB are piecewise linear curves connecting a finite number of data points.) This can be useful especially after zooming, because data points may then be spaced far apart. Another way to get coordinates of a point (in the Command Window, not the figure window) is by typing **ginput(1)**; this allows you to click on any point in the figure window, not just on a curve. While this gives you more flexibility, if you want the coordinates of a point on a curve, it is best to use the Data Cursor, because it will always highlight a point on the curve even if you don't click exactly on the curve.

The rightmost icon on the Figure Toolbar opens three windows surrounding the figure; these are collectively known as Plot Tools, and you can also open them with the command **plottools**. (These tools are not available in the Macintosh version of MATLAB 7.) You can also control the display of these three windows—the Figure Palette, the Plot Browser, and the Property Editor—individually from the **View menu**. These windows enable various means of editing figures, some of which we will discuss below. Many of these capabilities are also available in the **Insert** and **Tools** menus, and from the Plot Edit Toolbar in the

View menu.

You can use **subplot** to draw more than one graph within the same window, *e.g.*, to prepare side-by-side graphs for a printed figure. See the online help for a full explanation; as an example, the following commands draw the graph of $\sin x$ above the graph of $\cos x$, as shown in Figure 8.1:

```
>> subplot(2, 1, 1)
>> ezplot('sin(x)')
>> subplot(2, 1, 2)
>> ezplot('cos(x)')
```

You can also create multiple graphs within the same figure window using the **New Subplots** menu in the Figure Palette. Then select with the mouse the axes in which you want the next plot to appear.

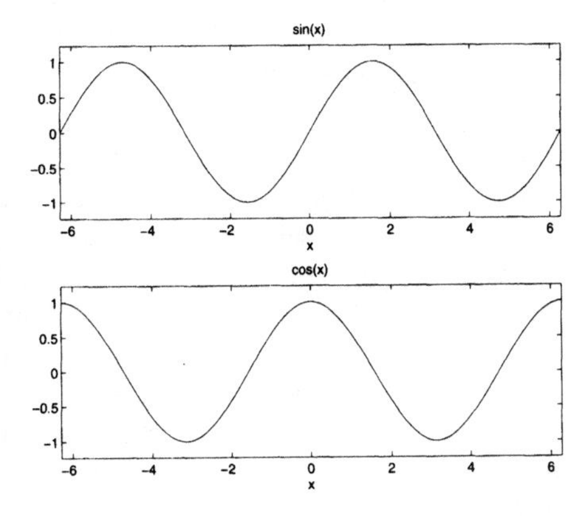

Figure 8.1: Two Graphs in one Figure Window

8.5.2 Editing Figures

In Section 3.8.2 we showed how to use **xlabel**, **ylabel**, and **title** to label axes and put a title on a graph. Here we describe several other ways to modify and annotate graphs.

When plotting more than one curve, it is often important to distinguish among them by making them have different colors or styles (*e.g.*, solid versus dashed). Once you've plotted a curve, you can change its appearance in the figure window as follows. First, click on the icon with the arrow, just to the right of the printer icon, in the Figure Toolbar. Then, right-click on the curve; the curve will be highlighted with black squares and a menu will appear. (If something else gets highlighted, try again.) You can change the color, width, or style of this curve, among other things, from this menu. Alternatively, if you have opened the Plot Tools windows surrounding the figure, as described above, you can select the curve by left-clicking on it, and modify it in the Property Editor. You can also use the Plot Browser to select a curve. Finally, when plotting with **plot** or **contour**, you can choose the line style or color in advance; for example, to plot the data in vectors **X** and **Y** as a dashed curve, type **plot(X, Y, '--')**. See the online help for **plot** for other options. When you have made each curve a different style or color, you can make a legend in the corner of the graph with the command **legend**; see its online help.

Another effective approach when presenting multiple curves on the same graph is to label the curves directly. You can add text, arrows, and other annotations using the **Insert** menu or, with Plot Tools enabled, using the Figure Palette. First select the type of annotation you want, then click the spot on the graph you want it to appear. For a text box, type the text next, while for an arrow, click again to indicate the other end of the arrow. If you don't want a box around your text, you can change its Edge Color to white by right-clicking on the box or by using the Property Editor. You can also add unboxed text by using the commands **text** and **gtext**; see their online help for details. Finally, to add a symbol exactly at specific coordinates—say, the coordinates of a special point you want to highlight on a curve—you can type **hold on** followed by an appropriate **plot** command, *e.g.*, **plot(1, 2, 'o')** to plot a small circle centered at the point $(1, 2)$. See the online help for **plot** for other available symbols.

You can change properties of a graph's axes, such as the font used to number the axes and label the graph using the Property Editor. First click on the axes or in a blank area within them to select them, much as you select a curve for editing. If the axes do not become highlighted, make sure the edit (arrow) icon in the Figure Toolbar is highlighted, and if not then click on this icon first. Then click on the Font tab in the Property Editor to change the font. Often a font that is larger than the default improves printed output. You can also change the font of a text annotation you have added by right-clicking on it and using the pop-up menu, or by selecting it and using the Property Editor.

To see all of the available ways to modify a curve or other object in the graph, the axes, or the entire figure window, select the object you want to modify and click Inspector in the Property Editor. (If the Property Editor is not open already, right-click on the object and select **Properties...**.) To select the figure window itself, click in the border outside the axes. Another way to see and change properties of these objects is with the commands **get** and **set**. To see the properties of the current figure, type **get(gcf)**; for the current axes, type **get(gca)**; and for the current selected curve or other object, type **get(gco)**. (The commands **gcf**, **gca**, and **gco** stand for "get current figure/axis/object".) To change, for example, the FontSize property of the current axes to 16, type **set(gca, 'FontSize', 16)**. While the the numbers on the axes will change to a 16 point font immediately, you will need to retype any previous **title**, **xlabel**, or **ylabel** commands to enlarge the title and labels.

8.6 Features of MATLAB's Numerical ODE Solvers

In Chapter 7 we introduced the command **ode45**, which solves initial value problems. Here we describe several additional ways to use this command. These features also apply to the other numerical solvers mentioned in Chapter 7, such as **ode15s**. Our discussion is based on the syntax available in MATLAB 6 (R12) and later versions, which is used only if the first input argument to **ode45** is a function handle. This is the case if you use **ode45** in the ways suggested in Chapter 7. But, if the first input argument of **ode45** is an inline function or a string containing the name of an M-file, it will use an old (MATLAB 5) syntax, and most of the features described below will not work as we describe them.

8.6.1 Evaluation of Numerical Solutions with deval

When solving the differential equation

$$\frac{dy}{dt} = f(t, y)$$

with **ode45**, we have seen that you can specify an interval of t values, in which case MATLAB computes the numerical solution at intermediate t values that it chooses, or you can specify a vector of t values, in which case MATLAB outputs the numerical solution at the t values you chose. Another approach is to let MATLAB generate the numerical solution at the intermediate t values it chooses and then use the command **deval** to evaluate (by interpolation) the numerical solution at whatever t values you choose.

In order to use **deval**, you must first assign the output of **ode45** to a single variable. In Chapter 7, we either assigned the output to two variables (to contain the computed t and y values) or did not assign the output at all (to immediately get a graph of the numerical solution). Let us instead type:

```
>> sol = ode45(@(t, y) t./y, [0 2], 1)
sol =
      solver:  'ode45'
     extdata:  [1x1 struct]
           x:  [1x11 double]
           y:  [1x11 double]
       stats:  [1x1 struct]
       idata:  [1x1 struct]
```

The output is then a structure that contains, among other things, the values of the numerical solution at various points in the specified t interval $[0, 2]$. These values are stored in the field **sol.y**, and the corresponding t values are stored in **sol.x**. (These fields are always called **x** and **y**, regardless of what letters you use for the independent and dependent variables). However, you need not concern yourself with these details in order to use **deval**.

Once you have stored the output of **ode45** into a structure, you can use **deval** to evaluate the numerical solution at any values of t within the interval given to **ode45**. For example:

```
>> deval(sol,0:0.5:2)
ans =
   1.0000    1.1180    1.4142    1.8028    2.2361
```

You can then run **deval** as many times as you want to get the numerical solution at other t values without having to re-run **ode45**.

You can use **deval** not only with the various other MATLAB commands that solve ordinary initial value problems, but also with the commands **bvp4c** (which we will use in Chapter 10 to solve boundary value problems), **dde23** (which solves delay differential equations), and **ode15i** (which solves implicit ODEs).

8.6.2 Plotting Families of Numerical Solutions of ODEs

In Chapter 7 we plotted a family of numerical solutions of a first order differential equation using a single **ode45** command and a vector of initial conditions. While this method works in many cases, it results in an incomplete graph if any one of the solutions cannot be computed over the entire specified interval. It is also unsuitable if the unknown function y is itself a vector, as will be the case when we use **ode45** in future chapters. In these cases, it is better to use a loop to compute and plot each curve separately.

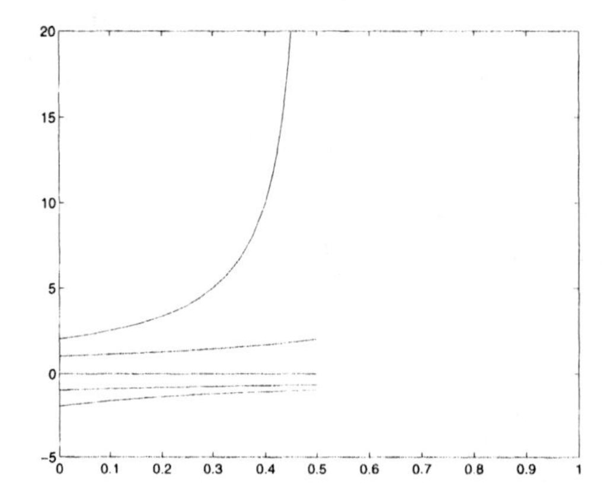

Figure 8.2: Solutions of $dy/dt = y^2$ up to the First Singularity

For example, suppose that you want to plot numerical solutions of

$$\frac{dy}{dt} = y^2$$

with $y(0) = -2, -1, 0, 1, 2$ on the interval $0 \leq t \leq 1$. You can produce Figure 8.2 by typing the following commands:

```
>> g = @(t, y) y.^2;
>> [t, y] = ode45(g, [0 1], -2:2);
Warning: Failure at t=4.999847e-001.    Unable to meet
integration tolerances without reducing the step size
below the smallest value allowed (8.881784e-016) at time t.
> In ode45 at 355
>> plot(t, y)
>> axis([0 1 -5 20])
```

(Notice that even though the right side of the differential equation is a function of y only, we have made **g** a function of two variables, as **ode45** requires.) Here **ode45** stops solving near $t = 0.5$ because that is where the solution with $y(0) = 2$ blows up to infinity, and all five solution curves are shown only for t between 0 and 0.5 even though four of them should extend further to the right. A better approach in this case is to define g and then use a loop as follows:

```
>> figure; hold on
>> for y0 = -2:2
       [t, y] = ode45(g, [0 1], y0);
       plot(t, y)
   end
>> hold off
>> axis([0 1 -5 20])
```

The result is in Figure 8.3.

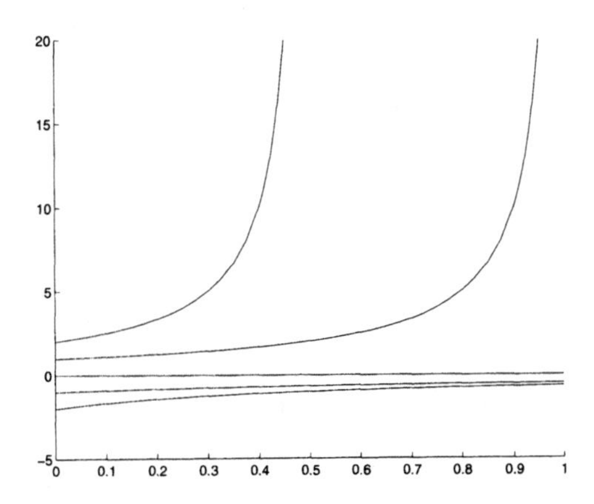

Figure 8.3: Solutions of $dy/dt = y^2$

In Chapters 10 and 13 we will discuss how to use **ode45** to numerically solve higher order equations and systems of equations. In these cases, the initial condition for each numerical solution will be a vector, and to plot a family of numerical solutions you must use a loop similar to the one above.

8.6.3 Event Detection

Sometimes the main question of interest in a problem that is modeled by a differential equation is when a specific event occurs; *e.g.*, when does a falling ball hit the ground? The MATLAB numerical ODE solvers have a feature that enables the detection of such events. Consider the first example from Chapter 7,

$$\frac{dy}{dt} = \frac{t}{y}, \qquad y(0) = 1,$$

and suppose you want to find the time when the solution y reaches the value 1.5. Since the exact solution of our initial value problem is $y(t) = \sqrt{t^2 + 1}$, we know that the desired value is $t = \sqrt{5}/2 = 1.118\ldots$. But for the sake of illustration, here are the steps to follow to approximate this value of t using **ode45**.

First, you must create an M-file, say `eventfunc.m`, that describes to **ode45** the event you want it to detect. Among other things, this function must specify a quantity that changes sign when the event occurs. Here is an appropriate event M-file for the problem described above:

```
function [event, stop, dir] = eventfunc(t, y)
event = y - 1.5;
stop = 1;
dir = 0;
```

Before explaining the details of this M-file, let us discuss how to use it to produce the desired output.

Once `eventfunc.m` is created, you can run the following commands:

```
>> options = odeset('Events', @eventfunc);
>> [t, y, tev, yev] = ode45(@(t,y) t./y, [0 2], 1, options);
>> [tev, yev]

ans =
     1.1180    1.5000
```

The first three input arguments to **ode45** are the usual ones. The time interval **[0 2]** should start at the time where the initial condition $y(0) = 1$ is given and end at a time by which you think the event will have occurred. We can see from Figure 7.1 that, in this example, y will indeed reach 1.5 before $t = 2$. The third and fourth outputs **tev** and **yev** from **ode45** are the t and y values at which the event occurs in the numerical solution. In the example above, the displayed digits of **tev** match those of the exact value $t = \sqrt{5}/2$, and of course **yev** is 1.5 for the event in our example. The first two outputs **t** and **y** are, as usual, vectors containing the intermediate values of t and y computed by **ode45**; they can be used for instance to plot the solution up to the time of the event.

You can use the event M-file and commands above as a template for many different problems. For other initial value problems and events, you need only change the time interval and initial condition you give to **ode45**, the line defining **event** in the M-file, and possibly the lines defining **stop** and **dir** (see below).

We conclude with a discussion of the specific format of the M-file `eventfunc.m` and the commands above. We turned on event detection in **ode45** by including a fourth argument **options** containing the output of the **odeset** command above (recall that we also used **odeset** in Section 7.3). With the property **Events** set to a function handle, **ode45** uses the given function to determine when an event occurs and what to do when it happens. This function must yield three outputs. The first output, which we called **event**, is a function of the solution y (and possibly of the independent variable t) that changes sign when (and only when) the event occurs. The second output, which we called **stop**, should be **1** if you want **ode45** to stop solving the differential equation once the event happens, or **0** if you want it to keep going and possibly detect multiple occurrences of the event.

The third argument, which we called **dir**, should be **1** if you want the event to occur only when **event** changes from negative to positive, **-1** for only positive to negative, or **0** for the event to occur in either direction. In the example above, we would have obtained the same answer with **dir = 1**, but no event would have been detected with **dir = -1**.

8.7 Troubleshooting

In this section, we list some advice for avoiding common mistakes. We also discuss warning messages and describe some techniques for recovering from errors.

8.7.1 The Most Common Mistakes

Here are some of the most common mistakes in using MATLAB.

1. Using `*`, `/`, or `^` instead of `.*`, `./`, or `.^` when trying to multiply, divide, or exponentiate vectors or matrices element-by-element. Unless you want to perform these operations by the rules of matrix multiplication, you should include the period before these three operands. In most cases you may include the period even when it is not necessary. However, you should not include the period in input to symbolic commands.

2. Using the wrong class of input to a command. Most of the commands described in this book use one or more of the following four classes of input: numeric, symbolic, character string, and function handle. Check the online help for a command if you are unsure what class of input it expects. If you use the wrong class of input, you may get an error message, or you may get a result that is different from what you intended.

3. Mismatching parentheses, brackets, quotes, or braces. MATLAB usually pinpoints the mistake, though the error message may be hard to interpret if you omit the quote at the end of a string.

4. Creating input or output lines that are too long. Though this does not interfere with the correct operation of MATLAB, it can make your printed output difficult or impossible to read. Depending on your method of printing, lines that are longer than about 70 characters will either be truncated or wrapped to the next line. For input lines, you can continue your input onto the next line by typing `...` followed by ENTER. Better still, try to separate long commands into shorter commands by using intermediate variables; see the *Sample Solutions* for examples. Sometimes long output lines are unavoidable, especially from the symbolic commands. In such cases, you can use the command **pretty** to format the output into shorter lines that are easier to read.

8.7.2 Error and Warning Messages

You should pay close attention to the error and warning messages that MATLAB generates. Although these messages may seem cryptic at first, you will find that they give valuable clues to locating mistakes in input. For example, suppose that you execute the following commands.

```
>> x = -1:0.01:2; plot(x, x^2)
??? Error using ==> mpower
Matrix must be square.
```

This error message points to the source of the problem, using ^ instead of **.^** with a vector. Be aware, however, that errors like this do not always produce error messages. For example, consider the following commands:

```
>> [x, y] = meshgrid(0:0.2:10, 0:0.2:10);
>> contour(x, y, x^2 + y^2)
```

Because **x** and **y** are square matrices, the use of ^ is allowed in this instance. However, because **.^** was not used, the resulting graph does not correctly depict level curves of $x^2 + y^2$.

Another common source of errors is using the wrong class of argument to a command. For example, suppose that you type

```
>> syms x; solve(x^2 - x = 1)
??? syms x; solve(x^2 - x = 1)
                          |
Error: The expression to the left of the equals sign is
not a valid target for an assignment.
```

Although the text of the error message is not easy to interpret, the vertical bar indicates where an error was first detected by MATLAB. In this instance, the problem is that the input argument cannot contain an equals sign unless it is a string, enclosed in single quotes. A trickier example is the following. Suppose that you defined, say, **x = 2** earlier in your session; then you type

```
>> int(sin(x))
??? Function 'int' is not defined for values of class
'double'.
```

This error message often arises when a symbolic function receives the wrong class of argument. In this case, since **x** is a floating point number, so is **sin(x)**, while **int** expects a string or symbolic expression. To correct the problem, type **syms x** and repeat the command, or enclose the argument in single quotes: **int('sin(x)')**.

It is important to distinguish between *error* messages and *warning* messages. An error message means that something is really wrong, and MATLAB probably will not be able to produce any output. A warning message is an indication that something might be wrong or that MATLAB can't do exactly what you've instructed it to do. Here is an example:

```
>> x = -5:0.01:5; plot(x, 1./x)
Warning: Divide by zero.
```

Although the vector **x** includes the value 0, at which $1/x$ cannot be evaluated, the commands above correctly graph $1/x$.

Chapter 9

Using Simulink

There is a useful auxiliary to MATLAB, called Simulink. Simulink is automatically included in the MATLAB Student Version; with the Professional Version, it can be obtained as an additional toolbox. In this chapter, we give a concise introduction to Simulink, focusing on its use to solve differential equations. Many students and engineers find that Simulink provides a convenient graphical "front end" to MATLAB's numerical solvers **`ode45`**, **`ode23`**, **`ode23s`**, *etc*. However, one should keep in mind that all the comments in the last chapter about reliability of numerical methods apply equally well to Simulink.

9.1 Constructing and Running a Simulink Model

We will illustrate the use of Simulink with one of the differential equations studied both in Chapter 6 and in Problem Set B, the nonlinear first-order differential equation (6.3): $y' = y^2 + t$ with the initial condition $y(0) = 0$. To launch Simulink, click on the icon on the Desktop toolbar, or else type **`simulink`** in the MATLAB Command Window. The first thing you will see is the Simulink Library, which looks like Figure 9.1 on *UNIX* systems or like Figure 9.2 on *Windows* machines.

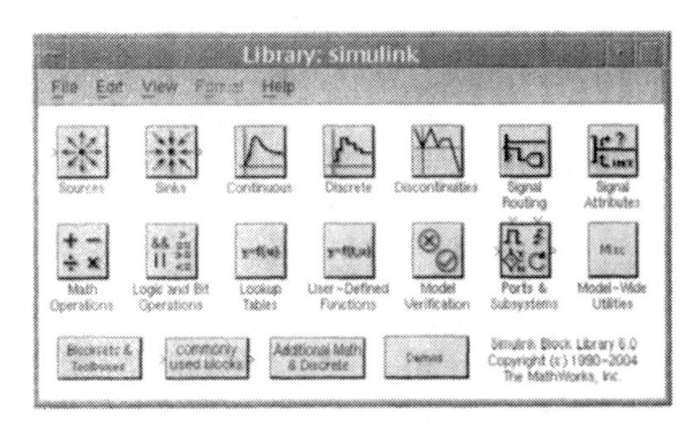

Figure 9.1: The Simulink Library

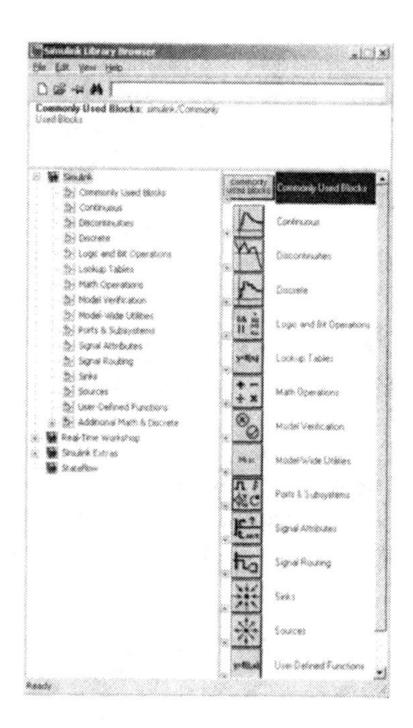

Figure 9.2: The Simulink Library Browser on a PC

To use Simulink to study our differential equation, we need to create a *model*, a graphical representation of the equation. The model results from thinking of a differential equation $y' = f(t, y)$ as a feedback diagram, where y and t are used to compute $f(t, y)$ and thus y', from which y is obtained by integrating, as in Figure 9.3. To begin to construct a model, first click on **New** and then **Model** from the **File** menu of either the MATLAB Desktop, the Simulink Library window, or the Editor/Debugger. A new *model window* will pop up. You create the model by dragging certain standard elements from sublibraries of the Simulink Library, called *blocks*, into the model window, and then connecting them together. For studying ordinary differential equations, most or all of the blocks you need will be found in the Sources Library, the Sinks Library, the Continuous Library, and the Math Operations Library. (See the list of Simulink blocks at the end of the Glossary for the locations of the most commonly used blocks for studying ordinary differential equations.)

One block you will almost always need is the Integrator block from the Continuous Library. In a first-order differential equation such as our sample equation $y' = y^2 + t$, the input to this block represents y', and the output to this block represents y. If you double-click on the Integrator block, a **Block Parameters** menu will pop up, in which there is a

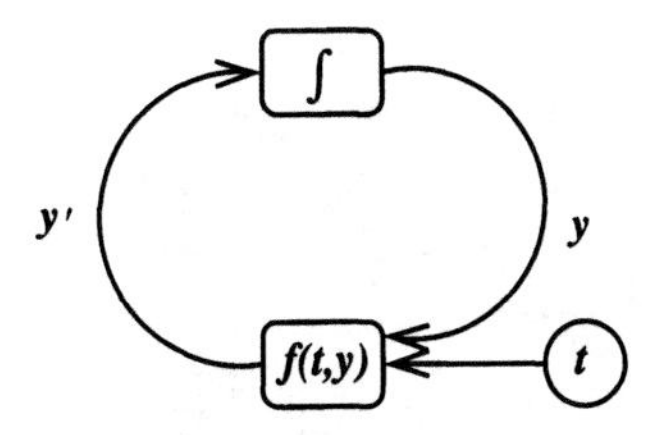

Figure 9.3: How to View $y' = f(t, y)$

place to put in an initial condition, which determines the "constant of integration." In our example, we don't have to change the default initial condition, which is $y(0) = 0$.

The model for our example, once finished, looks like Figure 9.4. Let's explain the

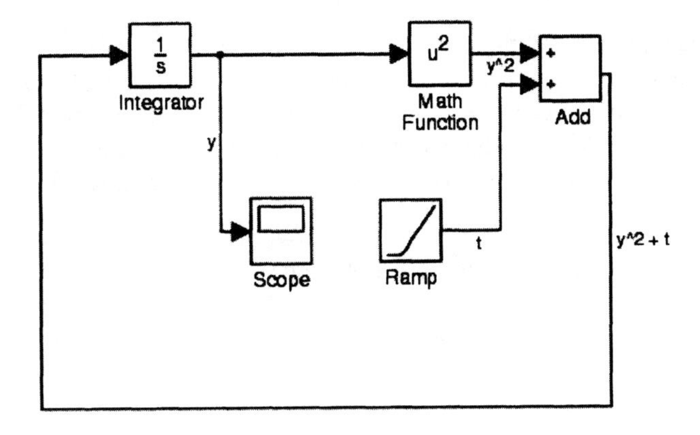

Figure 9.4: A Simulink Model for $y' = y^2 + t$

model, and then explain how to construct it from the blocks in the Simulink Library. The solid lines (with arrowheads on them) represent what Simulink calls *signals*; mathematically, these are variables. The blocks correspond to mathematical operations. Reading the model starting in the upper left, we see that the Integrator outputs the value of y. The Math Function block converts y to y^2, which is added to t coming from the Ramp block. (The Clock block, which outputs the value of t, could have also been used here.) Then the result, $y^2 + t$, goes back into the Integrator as the input y'. We read off the result with the Scope block, which draws a plot of y as a function of t. (Note that the Scope is attached to the signal y with a branch line.)

To give another example, Figure 9.5 shows a model for the linear equation $y' = (\sin t)y$. This model requires only four blocks.

You might wonder why we don't instead choose the Derivative block, and associate the input with y and the output with y'. While in principle this is possible, it doesn't usually

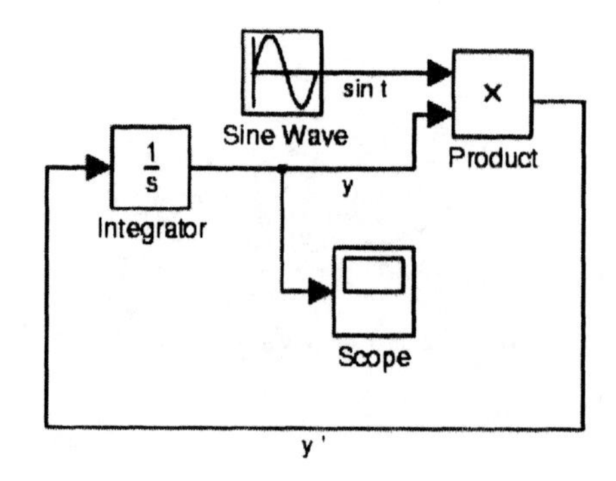

Figure 9.5: A Simulink Model for $y' = (\sin t)y$

yield good results, because of the fact that numerical differentiation is much less stable than numerical integration. (Keep in mind that Simulink does all its calculations numerically.) In fact, Figure 9.6 shows a second model for the same differential equation $y' = y^2 + t$, rewritten as $y = \sqrt{y' - t}$, with the use of the Derivative block. Later we will discuss the difference in results from the first model (Figure 9.4) and the second model (Figure 9.6).

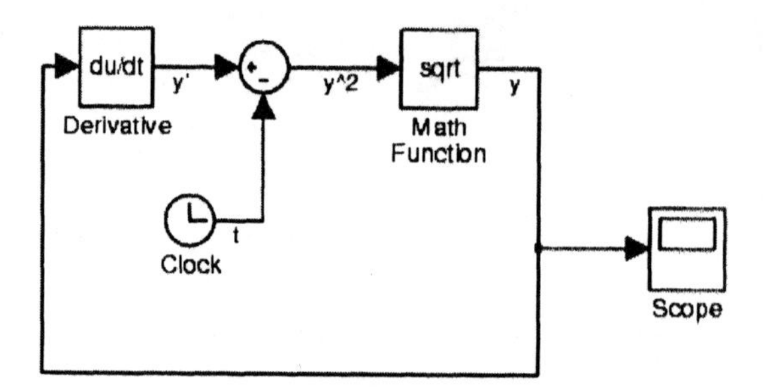

Figure 9.6: Another Simulink Model for $y' = y^2 + t$

First let's go back and discuss in detail how to construct the model in Figure 9.4. Start by double-clicking on the libraries you need in the Simulink Library window. Within each library, you will see a list of blocks represented by icons. Clicking on a block will display some help text on what it does. On *Windows* systems, for example, the Continuous Library, with the Integrator selected, looks like Figure 9.7.

Click on the blocks you need and drag them into the new model window. You don't need to worry so much about positioning at first, since you can rearrange things later. You will note that most blocks have one or more input ports and one or more output ports. (*Exception*: Blocks in the Sinks Library have no output ports; blocks in the Sources Library have no input ports.) Left click on a port and drag with the mouse to create a line going in

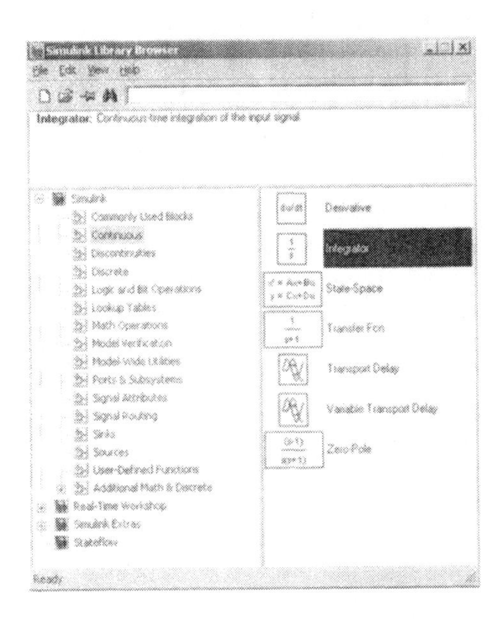

Figure 9.7: The Continuous Library

or out of that port. Usually you will want to drag the signal straight to a port on another block, but if it's necessary to bend the line, you may need to create the line in stages. The line will appear solid once it's connected at both ends. To create a branch line, hold down the CTRL key and left click the mouse at the point on an existing line where you want the branch line to start. You can rename a block or put a label on a signal (line) by right clicking on it, selecting **Block Properties...** or **Signal Properties...**, and then inserting the new name. Some blocks, such as the Sum block or the Math Function block, usually require customization. You do this (after you've dragged the block into your model window) by double-clicking on the block to call up the **Block Parameters** menu. For example, the default version of the Math Function block computes the exponential function and is labeled `e^u` accordingly, but there is a toggle switch to change the function to `log`, `square`, `sqrt`, *etc.* The default version of the Sum block has two input ports (at the "6:00" and "9:00" positions), each labeled with a + sign, but the **Block Parameters** menu calls up the List of Signs parameter. Setting this to $+ - +$, for example, means there are three inputs, and configures the block to subtract the second (middle) input from the first (top) input, and then add the third (bottom) input.

Once you have customized the blocks you need and connected them, you can slide them around with the mouse to make the appearance of your model as "clean" as possible.

Remember that to get useful output out of your model, you need to save room for at least one output block, usually a Scope block, that plots the signal it's connected to (in our case, y) as a function of time t.

When your model is complete, it's time for testing. This is the fun part! Hit **Save As...** from the **File** menu to give your model a name; the file extension `.mdl` (for "Simulink model") is added automatically. We called the example of Figure 9.4 `firstex.mdl`. Open the **Simulation** menu at the top of the model window and choose **Configuration Parameters...**. This enables you to set the initial and final values of t. For our example, a natural choice would be to keep the default starting value $t = 0$ and to change the "Stop Time" to $t = 2.0$. Now select **Start** (again from the **Simulation** menu). Don't be surprised if things do not go exactly as you planned and if you get various error messages. Usually such messages will point out silly mistakes, such as forgetting to set certain needed parameters or to connect all ports on some of the blocks. Make the necessary changes, hit **Save** to save them, and select **Start** again. You can also run your model from the Command Window using the MATLAB command **`sim('firstex')`**. Here the model name is enclosed in single quotes to make it into a string; one can add other arguments to **`sim`**, such as the range of values of t, but these other arguments to **`sim`** are optional. You can print a diagram of your model (like Figure 9.4) with the MATLAB command **`print -sfirstex`**, where **`firstex`** is of course replaced by whatever your model name happens to be.

Once your simulation does run successfully, it may appear that nothing has happened. This is not really the case. For one thing, it is likely that one or more new variables, created by the simulation, will show up in your Workspace. For example, when you run the model of Figure 9.4 with $0 \leq t \leq 2$, you will see a variable called **`tout`** in the Workspace, an array of size 109×1. This is a vector of the values of t where the (approximate) solution y has been computed. In this example, you should also get a warning message "Solver Step size is becoming less than specified minimum step size," that shows up in the Simulation Diagnostics window, and some other warning messages, saying "Warning: Unable to reduce the step size without violating minimum step size," in the MATLAB Command Window. Since we know that the solution blows up in finite time (in fact around $t = 1.986$), these warnings are not surprising. This example should convince you that sometimes Simulink's warning messages do not necessarily indicate that you made a mistake, but may be intrinsic to mathematical properties of your model. Finally, if you attached a Scope block to your output signal, the results of the simulation should be visible when you double click on the Scope block. The Scope window that pops up is not a figure window in the usual sense (it does not have a figure number, for example), but it does have many of the same features. With the buttons at the top, you can rescale the axes, adjust the tick marks on the axes, *etc.* For more details, look under Scope in the Help Browser window. The output of the model of Figure 9.4, once the axes have been suitably scaled, looks like Figure 9.8. This is very much like the graph of $y(t)$ that one obtains by solving the initial value problem with either **`dsolve`** or **`ode45`**.

Exercise 9.1 Build a Simulink model to study the differential equation $y' = y - 3e^{-2t}$, studied in Example 7.4. Vary the initial value of $y(0)$, and study how the numerical solution varies with the initial condition. Confirm what was shown in Figure 7.4, that slightly

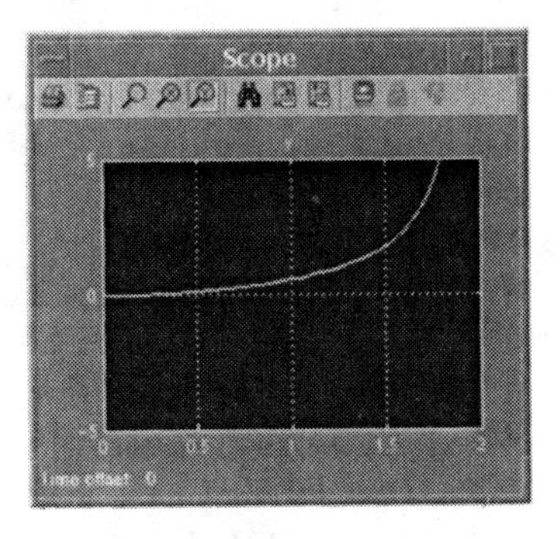

Figure 9.8: The Scope Output

different initial conditions give solutions which remain almost the same for a while, but eventually diverge from one another.

9.2 Output to the Workspace and How Simulink Works

To get more detailed information from Simulink, you may want to further process the output of Simulink using regular MATLAB commands. In order to do this, you should make use of the To Workspace block from the Sinks Library. For example, we could modify the model of Figure 9.4 by adding this block, as shown in Figure 9.9.

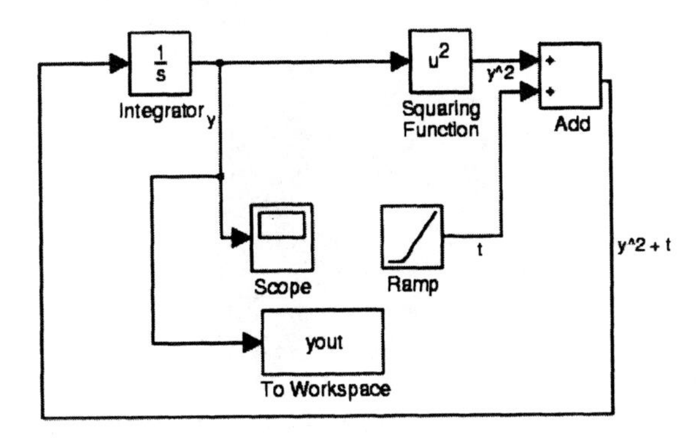

Figure 9.9: A Simulink Model with Workspace Output

The To Workspace block takes its input signal and outputs it to a vector with name specified in the **Block Parameters** menu. (By default this is called **`simout`** and is a structure;

in our example we've changed it to **yout**, and set it to be of class double array.) Running the model gives the same Scope output as before, but in addition leaves two arrays of size 109×1, **tout** and **yout**. These are just like the arrays of t and y values produced by **ode45**, and in fact the operation of Simulink depends on **ode45** running in the background. (Using the **Simulation** menu, you can change the default choice of numerical solver and use one of MATLAB's other solvers instead.)

Using the To Workspace block, we can now compare the models in Figure 9.4 and Figure 9.6. Running the model in Figure 9.6 again gives an error message in the Simulation Diagnostics window, this time saying: "Trouble solving algebraic loop containing `'secondex/Math Function'` at time 1.8. Stopping simulation. There may be a singularity in the solution." We also get a warning message in the Command Window, saying: "Warning: The model `'secondex'` does not have continuous states, hence using the solver `'VariableStepDiscrete'` instead of the solver `'ode45'` specified in the Configuration Parameters dialog." This is indicative of a slightly more serious problem than we had with the first model, and explains why using the Integrator block is better than using the Derivative block when you have a choice. For one thing, in the model of Figure 9.6, there is no convenient way to input the initial conditions, as there is with the Integrator block. Secondly, MATLAB needs to substitute a different solver for **ode45**. Thirdly, there is the problem that the square root function is not differentiable at 0 and not defined to the left of 0. And finally, there is the issue we already discussed, namely that numerical differentiation is more unstable than numerical integration. In spite of all these problems, the simulation still "works"—the picture one sees in the Scope is shown in Figure 9.10, and very much resembles Figure 9.8.

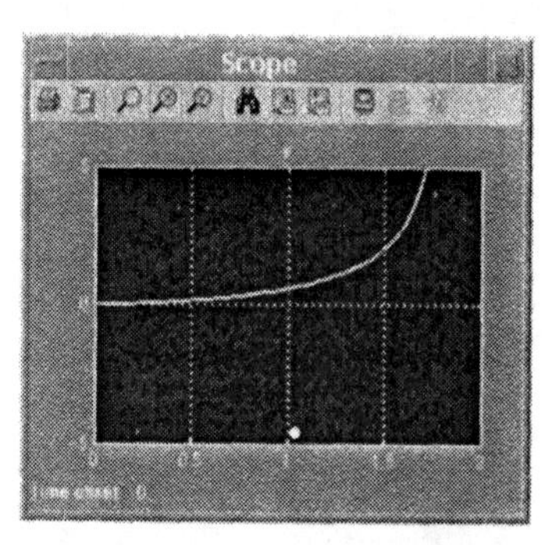

Figure 9.10: Scope Output for the Second Model

If we output the result of the second model to the Workspace as we did with the first model, we can now compare the results. The final results are shown in Figure 9.11. The computed points from the first model are shown with diamonds; those from the second model are shown with circles. Note that the two solutions are almost identical out to $t =$

1.5, but then they start to diverge. The exact solution to the equation, as computed with **dsolve**, is shown with a solid line, and basically coincides with the results of the first model.

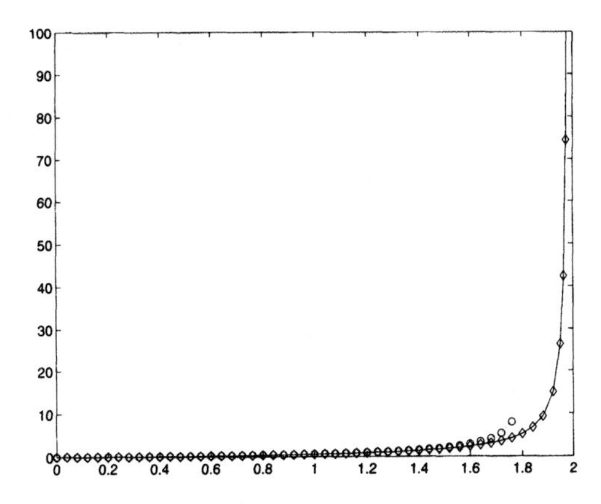

Figure 9.11: Comparing Two Models

Exercise 9.2 Build a Simulink model to study the nonlinear (autonomous) initial value problem

$$\frac{dy}{dt} = \cos y, \qquad y(0) = 0, \tag{9.1}$$

and to output the approximate solution to the Workspace. Compare your results with the exact solution as computed by **dsolve**. Confirm that y approaches $\frac{\pi}{2}$ as t gets large.

Problem Set C

Numerical Solutions

In this problem set you will use **`ode45`** and **`plot`** to calculate and plot numerical solutions to ordinary differential equations. The use of these commands is explained in Chapters 3, 7, and 8. The solution to Problem 3 appears in the *Sample Solutions*.

1. We are interested in describing the solution $y = \phi(t)$ to the initial value problem

$$\frac{dy}{dt} = 2\frac{t}{y} + e^y, \qquad y(0) = 1. \tag{C.1}$$

Observe that $dy/dt \geq e^y \geq e$ in the region $t \geq 0, y \geq 1$, and therefore $\phi(t)$ increases to ∞ as t increases. But how fast does it increase?

(a) First of all, $\phi(t)$ must increase at least as fast as the solution $y = \phi_0(t)$ to the initial value problem

$$\frac{dy}{dt} = e^y, \qquad y(0) = 1.$$

Solve this problem symbolically and conclude that $\phi(t) \to \infty$ as $t \to t^*$ for some $t^* \leq 1/e < 1/2$.

(b) Next, since $2t/y \leq 1$ for $t \leq 1/2$ and $y \geq 1$, it follows that $\phi(t)$ increases at most as fast as the solution $y = \phi_1(t)$ to the initial value problem

$$\frac{dy}{dt} = 1 + e^y, \qquad y(0) = 1.$$

Solve this problem symbolically and conclude that $t^* \geq \ln(e + 1) - 1$.

(c) Now compute a numerical solution of (C.1) and find an approximate value of t^*.

(d) Finally, plot the numerical solution on the same graph with $\phi_0(t)$ and $\phi_1(t)$ and compare the solutions.

2. Consider the initial value problem

$$y' = y^2 - e^{-t^2}, \qquad y(0) = 2. \tag{C.2}$$

Since **dsolve** is unable to solve this equation, we need to use qualitative and numerical methods. Observe that $0 < e^{-t^2} \le 1$, so that

$$y^2 \ge y^2 - e^{-t^2} \ge y^2 - 1.$$

Therefore the solution $\phi(t)$ to (C.2) must satisfy $\phi_0(t) \ge \phi(t) \ge \phi_1(t)$, where ϕ_0 and ϕ_1 are the solutions to the respective initial value problems

$$\begin{cases} y' = y^2 & y(0) = 2, \\ y' = y^2 - 1 & y(0) = 2. \end{cases}$$

   (a) Solve for ϕ_0 and ϕ_1 explicitly (using **dsolve**) and conclude that $\phi(t) \to \infty$ as $t \to t^*$, for some $t^* \in [0.5, 0.5 \ln 3]$.

   (b) Compute a numerical solution of (C.2), find an approximate value of t^*, and plot $\phi(t)$ for $0 \le t < t^*$.

3. We shall study solutions $y = \phi_b(t)$ to the initial value problem

$$y' = (y - t)(1 - y^3), \qquad y(0) = b$$

for nonnegative values of t.

   (a) Plot numerical solutions $\phi_b(t)$ for several values of b. Include values of b that are less than or equal to 0, between 0 and 1, equal to 1, and greater than 1.

   (b) Now, based on these plots, describe the behavior of the solution curves $\phi_b(t)$ for positive t, when $b \le 0, 0 < b < 1, b = 1$, and $b > 1$. Identify limiting behavior and indicate where the solutions are increasing or decreasing.

   (c) Next, combine your plots with a plot of the line $y = t$. The graph should suggest that the solution curves for $b > 1$ are asymptotic to this line. Explain from the differential equation why that is plausible. (*Hint*: Use the differential equation to consider the sign of y' on and close to the line $y = t$.)

   (d) Finally, superimpose a plot of the direction field of the differential equation to confirm your analysis.

4. We shall study solutions $y = \phi_b(t)$ to the initial value problem

$$y' = (y - \sqrt{t})(1 - y^2), \qquad y(0) = b$$

for nonnegative values of t.

   (a) Plot numerical solutions $\phi_b(t)$ for several values of b. Include values of b that are less than -1, equal to -1, between -1 and 1, equal to 1, and greater than 1.

(b) Now, based on these plots, describe the behavior of the solution curves $\phi_b(t)$ for positive t, when $b < -1$, $b = -1$, $-1 < b < 1$, $b = 1$, and $b > 1$. Identify limiting behavior and indicate where the solutions are increasing or decreasing.

(c) By combining your plots with a plot of the parabola $y = \sqrt{t}$, show that the solution curves for $b > 1$ are asymptotic to this parabola. Explain from the differential equation why that is plausible.

(d) Finally, superimpose a plot of the direction field of the differential equation to confirm your analysis.

5. In this problem, we analyze the Gompertz-threshold model from Problem 17 of Problem Set B. That is, consider the differential equation

$$y' = y(1 - \ln y)(y - 3).$$

Using various nonnegative values of $y(0)$, find and plot several numerical solutions on the interval $0 \leq t \leq 6$. By examining the differential equation and analyzing your plots, identify all equilibrium solutions and discuss their stability.

6. Consider the initial value problem

$$\frac{dy}{dt} = \frac{t - e^{-t}}{y + e^{y}}, \qquad y(1.5) = 0.5.$$

(a) Use **ode45** to find approximate values of the solution at $t = 0, 1, 1.8$, and 2.1. Then plot the solution.

(b) In this part you should use the results from parts (c) and (d) of Problem 5 in Problem Set B (which appears in the *Sample Solutions*). Compare the values of the actual solution and the numerical solution at the four specified points. Plot the actual solution and the numerical solution on the same graph.

(c) Now plot the numerical solution on several large intervals (*e.g.*, $1.5 \leq t \leq 10$ or $1.5 \leq t \leq 100$). Make a guess about the nature of the solution as $t \to \infty$. Try to justify your guess on the basis of the differential equation.

7. Consider the initial value problem

$$e^{y} + (te^{y} - \sin y)\frac{dy}{dt} = 0, \qquad y(2) = 1.5.$$

(a) Use **ode45** to find approximate values of the solution at $x = 1, 1.5$, and 3. Then plot the solution on the interval $0.5 \leq t \leq 4$.

(b) In this part you should use the results from parts (c) and (d) of Problem 6 in Problem Set B. Compare the values of the actual solution and the numerical solution at the three specified points. Plot the actual solution and the numerical solution on the same graph.

(c) Now plot the numerical solution on several large intervals (*e.g.*, $2 \leq t \leq 10$, or $2 \leq t \leq 100$, or even $2 \leq t \leq 1000$). Make a guess about the nature of the solution as $t \to \infty$. Try to justify your guess on the basis of the differential equation. Similarly, plot the numerical solution on intervals $\epsilon \leq t \leq 2$ for different choices of ϵ as a small positive number. What do you think is happening to the solution as $t \to 0+$?

8. The differential equation

$$y' = e^{-t^2}$$

cannot be solved in terms of elementary functions, but it can be solved with **dsolve**.

(a) Use **dsolve** to solve this equation.

(b) The answer is in terms of **erf**, a special function (*cf.* Chapter 5), known as the *error function*. Use the MATLAB differentiation operator **diff** to see that

$$\frac{d}{dx}(\text{erf}(t)) = \frac{2}{\sqrt{\pi}}e^{-t^2}.$$

In fact,

$$\text{erf}(t) = \frac{2}{\sqrt{\pi}}\int_0^x e^{-s^2}\,dt.$$

Use **help erf** to confirm this formula.

(c) Although one does not have an elementary formula for this function, the numerical capabilities of MATLAB mean that we "know" this function as well as we "know" elementary functions like $\tan t$. To illustrate this, evaluate $\text{erf}(t)$ at $t = 0, 1$, and 10.5, and plot $\text{erf}(t)$ on $-10 \leq t \leq 10$.

(d) Compute $\lim_{t\to\infty} \text{erf}(t)$ and $\int_{-\infty}^{\infty} e^{-t^2}\,dt$.

(e) Next, solve the initial value problem

$$y' = 1 - 2ty, \qquad y(0) = 0$$

using **dsolve**. You should have gotten a formula involving **erf** with a complex argument. Solve this initial value problem by hand-calculation. Your answer will involve an integral that cannot be evaluated by elementary techniques. Since solutions of initial value problems are unique, these two formulas (the one obtained by MATLAB and the one obtained by hand-calculation) must be the same. Explain how they are related. Hint: Use **int** to evaluate $\int_0^t e^{s^2}\,ds$.

9. Consider the initial value problem

$$ty' + (\sin t)y = 0, \qquad y(0) = 1.$$

(a) Use **dsolve** to solve the initial value problem.

(b) Note the occurrence of the built-in function **sinint**, called the *Sine Integral* function and usually written $\mathrm{Si}(t)$. Check that this function is an antiderivative of $\sin t/t$ by differentiating it.

(c) Evaluate $\lim_{t\to\infty} \mathrm{Si}(t)$. Plot $\mathrm{Si}(t)$ and discuss the features of the graph.

(d) Do the same with the solution to the initial value problem.

(e) Now solve the initial value problem using **ode45**, and plot the computed solution. Compare your plot to the one obtained in part (d). You will find that MATLAB cannot evaluate $\sin t/t$ at $t = 0$, even though the singularity is removable. One way to get around this is to give the initial condition $y = 1$ at a value of t extremely close to, but not equal to, zero. Since we are only finding an approximate solution with **ode45** anyhow, there shouldn't be much harm done as long as the amount by which we move the initial condition is small compared with the error we expect from the numerical procedure.

(f) Discuss the stability of the differential equation. Illustrate your conclusions by graphing solutions with different initial values on the interval $[-10, 10]$.

Note: In this problem some of your plots may take a long time to generate.

10. Solve the following initial value problems numerically, then plot the solutions. Based on your plots, predict what happens to each solution as t increases. In particular, if there is a limiting value for y, either finite or infinite, find it. If it is unclear from the plot you've made, try replotting on a larger interval. Another possibility is that the solution blows up in finite time. If so, estimate the time when the solution blows up. Try to use the qualitative methods of Chapter 6 to confirm your answers.

(a) $y' = e^{-3t} + \dfrac{1}{1+y^2}, \quad y(0) = -1.$

(b) $y' = e^{-2t} + y^2, \quad y(0) = 1.$

(c) $y' = \cos t - y^3, \quad y(0) = 0.$

(d) $y' = (\sin t)y - y^2, \quad y(0) = 2.$

11. Solve the following initial value problems numerically, then plot the solutions. Based on your plots, predict what happens to each solution as t increases. See Problem 10 for additional instructions.

(a) $y' = 3t - 5\sqrt{y}, \quad y(0) = 4.$

(b) $y' = (t^3 - y^3)\cos y, \quad y(0) = -1.$

(c) $y' = \dfrac{y^2 + 2ty}{t^2 + 3}, \quad y(1) = 2.$

(d) $y' = -2t + e^{-ty}, \quad y(0) = 1.$

12. Construct a Simulink model to represent the initial value problem

$$y' = \sin y + C \sin t, \qquad y(0) = 0.$$

(*Hint*: For the forcing term $C \sin t$, use the Sine Wave block. When you left-click on this block to bring up the **Block Parameters** window, you can enter the amplitude C in the appropriate box.) Run your model for $0 \leq t \leq 35$ for various values of C (starting with $C = 1$ and $C = -1$) and see what you observe. You can run the model with many values of C simultaneously by entering a vector value for the Sine Wave amplitude. Set the amplitude to the vector `-4:4` and observe the results of the simulation in a Scope window. How can you interpret the results in view of an analysis of the equilibrium solutions of the autonomous equation $y' = \sin y$?

13. This problem illustrates one of the possible pitfalls of blindly applying numerical methods without paying attention to the theoretical aspects of the differential equation itself. Consider the equation

$$ty' + 3y - 9t^2 = 0.$$

(a) Use the MATLAB program in Chapter 7 to compute the Euler Method approximation to the solution with initial condition $y(-0.5) = 3.15$, using step size $h = 0.2$ and $n = 10$ steps. The program will generate a list of ordered pairs (x_i, y_i). Use **`plot`** to graph the piecewise linear function connecting the points (x_i, y_i).

(b) Now modify the program to implement the Improved Euler Method. Can you make sense of your answers?

(c) Next, use **`ode45`** to find an approximate solution on the interval $(-0.5, 0.5)$, and plot it with **`plot`**. Print out the values of the solution at the points **`-0.06:0.02:0.06`**. What is the interval on which the approximate solution is defined?

(d) Solve the equation explicitly and graph the solutions for the initial conditions $y(0) = 0$, $y(-0.5) = 3.15$, $y(0.5) = 3.15$, $y(-0.5) = -3.45$, and $y(0.5) = -3.45$. Now explain your results in (a)–(c). Could we have known, without solving the equation, whether to expect meaningful results in parts (a) and (b)? Why? Can you explain how **`ode45`** avoids making the same mistake?

14. Consider the initial value problem

$$\frac{dy}{dt} = e^{-t} - 3y, \qquad y(-1) = 0.$$

(a) Use the MATLAB program `myeuler.m` from Chapter 7 to compute the Euler Method approximation to $y(t)$ with step size $h = 0.5$ and $n = 4$ steps. The program will generate a list of ordered pairs (t_i, y_i). Use **`plot`** to graph the piecewise linear function connecting the points (t_i, y_i). Repeat with $h = 0.2$ and $n = 10$.

(b) Now modify the program to implement the Improved Euler Method, and repeat part (a) using the modified program. Find the exact solution of the initial value problem and plot it, the two Euler approximations, and the two Improved Euler approximations on the same graph. Label the five curves.

(c) Now use the Euler Method program with $h = 0.5$ to approximate the solution on the interval $[-1, 9]$. Plot both the approximate and exact solutions on this interval. How close is the approximation to the exact solution as t increases? In light of the discussion of stability in Chapters 5 and 7, explain your results in parts (a)–(c).

15. Consider the initial value problem

$$\frac{dy}{dt} = 2y + \cos t, \qquad y(0) = -2/5.$$

(a) Use the MATLAB program in Chapter 7 to compute the Euler Method approximation to $y(t)$ with step size $h = 0.5$ and $n = 12$ steps. The program will generate a list of ordered pairs (t_i, y_i). You do not need to print out these numbers, but graph the piecewise linear function connecting the points (t_i, y_i). What appears to be happening to y as t increases?

(b) Repeat part (a) with $h = 0.2$ and $n = 30$, and then with $h = 0.1$ and $n = 60$, each time plotting the results on the same set of axes as before. How are the approximate solutions changing as the step size decreases? Can you make a reliable prediction about the long-term behavior of the solution?

(c) Use **ode45** to find an approximate solution and again plot it on the interval $[0, 6]$ on the same set of axes as before. Now what does it look like y is doing as t increases? Next, plot the solution from **ode45** on a larger interval (going to $t = 15$) on a new set of axes. Again, what is happening to y as t increases?

(d) Solve the initial value problem exactly and compare the exact solution to the approximations found above. In light of the discussion of stability in Chapters 5 and 7, explain your results in parts (a)–(c).

16. Consider the initial value problem

$$\frac{dy}{dx} = 2y - 2 + 3e^{-t}, \qquad y(0) = 0.$$

(a) Use the MATLAB program in Chapter 7 to compute the Euler Method approximation to $y(t)$ with step size $h = 0.2$ and $n = 10$ steps. The program will generate a list of ordered pairs (t_i, y_i). Use **plot** to graph the piecewise linear function connecting the points (t_i, y_i). What appears to be happening to y as t increases?

(b) Repeat part (a) with $h = 0.1$ and $n = 20$, and then with $h = 0.05$ and $n = 40$. How are the approximate solutions changing as the step size decreases? Can you make a reliable prediction about the long-term behavior of the solution?

(c) Use **ode45** to find an approximate solution and plot it on the interval $[0, 2]$. Now what does it look like y is doing as t increases? Next, plot the solution from **ode45** on a larger interval (going at least to $t = 10$). Again, what is happening to y as t increases?

(d) Solve the initial value problem exactly and compare the exact solution to the approximations found above. In light of the discussion of stability in Chapters 5 and 7, explain your results in parts (a)–(c).

17. The function **erf**, discussed in Chapter 5 and in Problem 8 in this set, is the solution to the initial value problem

$$\frac{dy}{dt} = \frac{2}{\sqrt{\pi}} e^{-t^2}, \qquad y(0) = 0,$$

so if we solve this initial value problem numerically we get approximate values for the built-in function **erf**. Use **ode45**, employing the accuracy options discussed in Chapter 7, to calculate values for $\text{erf}(0.1), \text{erf}(0.2), \ldots, \text{erf}(1)$ having at least 10 correct digits. Present your results in a table. In a second column print the values of $\text{erf}(x)$ for $x = 0.1, 0.2, \ldots, 1$, obtained by using the built-in function **erf**. Compare the two columns of values.

18. Consider the Gompertz-threshold model,

$$y' = y(1 - \log y)(y - 3).$$

From the direction field and the qualitative approach for this equation, one learns that the solution with initial condition, $y(0) = 5$, approaches 3 as t approaches ∞. Use the event detection feature of **ode45** to find the time t at which $y = 3.1$.

19. Build a Simulink model for the initial value problem

$$y' = y^3 + t, \quad y(0) = 0,$$

which cannot be solved by **dsolve**. Display the graph of $y(t)$ in a Scope block.

(a) When you run the model with the default parameters (t going from 0 to 10), what error message do you get? What do you think is responsible for this, mathematically?

(b) The solution to this IVP only exists for $t < t^*$, for some positive number t^*. What does the model tell you about the (approximate) value of this number?

(c) How does the graph of $y(t)$ for $0 \le t < t^*$ (as displayed in the Scope window) compare with what you expected?

Chapter 10

Solving and Analyzing Second Order Linear Equations

Newton's second law of dynamics—force is equal to mass times acceleration—tells physicists that, in order to understand how the world works, they must pay attention to forces. Since acceleration is a second derivative, the law also tells us that second order differential equations are likely to appear when we apply mathematics to study the real world.

We note that in Chapters 5, 6 and 7 we discussed three different approaches to solving first order differential equations: searching for exact formula solutions; using geometric methods to study qualitative properties of solutions—typically when we cannot find a formula solution; and invoking numerical methods to produce approximate solutions—again when no solution formula is attainable.

In this chapter we shall bring each of these methods to bear on second order equations. The most basic second order differential equations are linear equations with *constant coefficients*:

$$ay'' + by' + cy = g(t).$$

These equations model a wide variety of physical situations, including oscillations of springs, simple electric circuits, and the vibrations of tuning forks to produce sound and of electrons to produce light. In other situations, such as the motion of a pendulum, we may be able to approximate the resulting differential equation reasonably well by a linear differential equation with constant coefficients. Fortunately, we know (and MATLAB can apply) several techniques for finding explicit solution formulas to linear differential equations with constant coefficients.

Unfortunately, we often cannot find solution formulas for more general second order equations, even for linear equations with *variable coefficients*:

$$y'' + p(t)y' + q(t)y = g(t).$$

Such equations have important applications to physics. For example, Airy's equation,

$$y'' - ty = 0, \tag{10.1}$$

arises in diffraction problems in optics, and Bessel's equation,

$$y'' + \frac{1}{t}y' + y = 0, \tag{10.2}$$

occurs in the study of vibrations of a circular membrane and of water waves with circular symmetry.

When studying first order differential equations for which exact solution formulas were unavailable, we could turn to several other methods. By specifying an initial value, we could compute an approximate numerical solution. By letting the initial values vary, we could plot a one-parameter family of approximate solution curves and get a feel for the behavior of a general solution. Alternatively, we could plot the direction field of the differential equation and use it to draw conclusions about the qualitative behavior of solutions.

For second order equations, these methods are more cumbersome; you must specify two conditions to pick out a unique solution. We usually specify initial values for the function and its first derivative at some point. Then we can use numerical techniques to compute an approximate solution. In order to graph enough solutions to get an idea of their general behavior, we must construct a two-parameter family of solutions. Since the initial value does not determine the initial slope, we cannot draw a direction field for a second order equation.

In some applications, the differential equation is accompanied not by initial conditions, but by *boundary conditions*; *i.e.*, we specify values $y(t_0) = y_0$ and $y(t_1) = y_1$ at two distinct points. The resulting problem is called a *boundary value problem*. We shall experiment with a new MATLAB command **bvp4c** that can solve some boundary value problems.

In this chapter, we describe how to use MATLAB to find formula solutions of second order linear differential equations with constant coefficients. We also describe how to find and plot numerical solutions to more general second order differential equations. In addition, we describe a method for solving boundary value problems using MATLAB's numerical solver. Finally, we introduce comparison methods and a more sophisticated geometric method for analyzing second order linear equations with variable coefficients. These qualitative methods have an advantage over the more obvious numerical and graphical methods. They more effectively yield information about properties shared by all solutions of a differential equation, about the oscillatory nature of solutions of an equation, and about the precise rate of decay or growth of solutions.

Comparison methods provide information on the solutions of a variable coefficient equation by comparing the equation with an appropriate constant coefficient equation. This possibility is suggested by the following example. For large t, the coefficient $1/t$ in Bessel's equation (10.2) is close to 0. So, Bessel's equation is close to the equation $y'' + y = 0$, whose general solution can be written as $y = R\cos(t-\delta)$. Thus one might expect solutions to Bessel's equation to oscillate and look roughly like sine waves for large t. We present a result that validates such comparisons.

The geometric method is based on direction fields, which are not directly applicable to a second order equation. We show, however, how to construct a related first order equation whose direction field yields information about the solutions of the original second order equation.

10.1 Second Order Equations with MATLAB

MATLAB's usual tools for finding symbolic solutions of differential equations work perfectly well for second order linear differential equations with constant coefficients. The syntax of the command is familiar, except that we must specify an additional initial condition.

Example 10.1 Consider the differential equation

$$y'' + y' - 6y = 20e^t. \tag{10.3}$$

We can get a solution by typing:

```
>> ode1 = 'D2y + Dy - 6*y = 20*exp(t)';
>> dsolve(ode1)
ans =
exp(2*t)*C2+exp(-3*t)*C1-5*exp(t)
```

As expected, the general solution depends on two arbitrary constants. To solve the differential equation with initial conditions $y(0) = 0$ and $y'(0) = 1$, we would type

```
>> dsolve(ode1, 'y(0) = 0', 'Dy(0) = 1')
ans =
4/5*exp(2*t)+21/5*exp(-3*t)-5*exp(t)
```

MATLAB can also solve some boundary value problems. Let's continue using the same differential equation, and consider the boundary conditions $y(0) = 0$ and $y(\ln(2)) = 10$. We get the solution of the boundary value problem by typing:

```
>> dsolve(ode1, 'y(0) = 0', 'y(log(2)) = 10')
ans =
5*exp(2*t)-5*exp(t)
```

Example 10.2 Now consider the linear second order differential equation

$$y'' + t^2y' + y = 0. \tag{10.4}$$

MATLAB is unable to find an explicit formula for the general solution. Instead, we can use **ode45** to find a numerical solution. But **ode45** is designed only for first order equations. Fortunately, it can deal with vector, as well as scalar, equations. Therefore, we convert the second order scalar equation to a first order vector equation for **y**, with **y(1)** representing the y-value and **y(2)** representing the value of its first derivative. Thus to solve a second order differential equation numerically in MATLAB, we must rewrite it in the form $\frac{d}{dt}[y; y'] = [y'; f(t, y, y')]$, and translate the formula for $[y'; f(t, y, y')]$ into an anonymous function (or an inline function if you are using MATLAB R13 or earlier). To illustrate how this works we recast (10.4) into $y'' = -t^2y' - y$, and define

```
>> rhs = @(t, y) [y(2); -t^2*y(2) - y(1)];
```

Now the numerical solution defined on the interval $[0, 3]$ with initial conditions $y(0) = 1$, $y'(0) = -1$ can be found by typing

```
>> [xa, ya] = ode45(rhs, [0 3], [1 -1]);
```

and it can be plotted using the techniques developed in Chapter 7.

A good way to understand the behavior of a general solution of a second order equation is to plot numerical solutions corresponding to a wide range of initial values and initial slopes. Here are the MATLAB commands to do so for this example.

```
>> figure, hold on
>> for y0 = -2:2
      for yp0 = -1:0.5:1
         [tfor, yfor] = ode45(rhs, [0  3], [y0 yp0]);
         [tbak, ybak] = ode45(rhs, [0 -2], [y0 yp0]);
         plot(tfor, yfor(:,1))
         plot(tbak, ybak(:,1))
      end
   end
```

The result is shown in Figure 10.1. We see that the solutions seem to level off for positive t and seem to blow up for negative t.

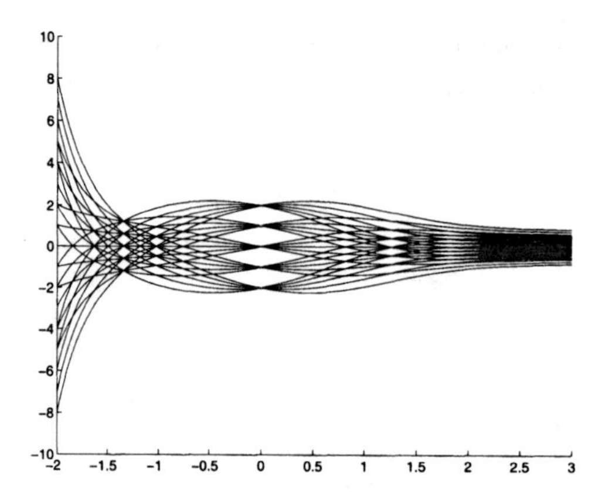

Figure 10.1: Numerical Solutions of (10.4)

Could we have obtained this qualitative information directly from the differential equation instead of from the graph of a few solutions? One approach is to look at the solutions of differential equations that are similar to this one. For example, when t is small, t^2 is close to zero. So the solutions to (1.4) should be close to the solutions of the differential equation

$$y'' + y = 0.$$

We know that the solutions to the latter equation are sine curves. For example, the solution satisfying the initial conditions $y(0) = 0$, $y'(0) = 1$ is just $\sin(t)$. So, one might expect the solutions of this initial value problem to look like $\sin(t)$ for small values of t. In

Figure 10.2, we have plotted $\sin(t)$ (with a dashed line) and the numerical solution to the initial value problem on the same axes.

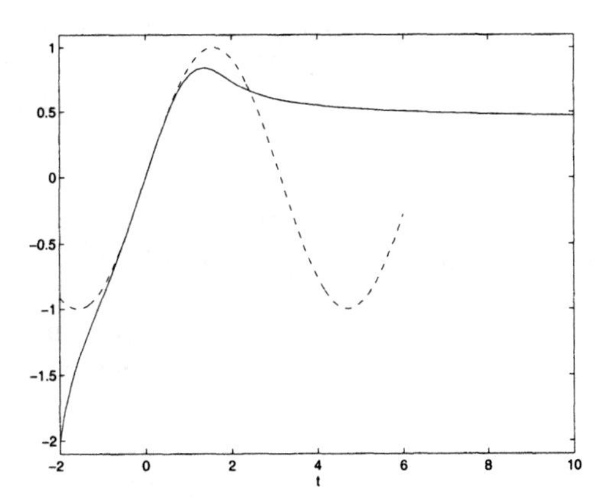

Figure 10.2: Comparing (10.4) with $y'' + y = 0$

The graph of the numerical solution in Figure 10.2 appears to level off as t increases. Looking again at the differential equation, we see that when t is large, the t^2y' term should dominate the y term. So, the solutions to the differential equation should be close to those of

$$y'' + t^2y' = 0.$$

This equation can almost be solved explicitly; its general solution is given by

$$y(t) = A + B\int_0^t e^{-u^3/3}\,du.$$

The exponential function being integrated is nearly zero for large positive values of u. So, when t is large and positive, the solutions should become almost constant. When t is large and negative, the exponential term is huge, and we expect the solutions to blow up.

Finally, let's look at the boundary value problem

$$y'' + t^2y' + y = 0, \quad y(0) = 0, \quad y(1) = 1. \tag{10.5}$$

Here we have the same differential equation, but we have specified values of the solution at two different points; $y(0) = 0$ and $y(1) = 1$ are called *boundary conditions*. This time **dsolve** cannot find a solution, so we discuss two alternate approaches. The first is called the *shooting method*. Consider the initial value problem

$$y'' + t^2y' + y = 0, \quad y(0) = 0, \quad y'(0) = s, \tag{10.6}$$

where the first initial condition is the first boundary condition, and in the second initial condition, the slope at $t = 0$ is specified by the parameter s. Since for any value of s, the solution of this initial value problem satisfies the first boundary condition in (10.5), if we find the value $s = \bar{s}$ for which the solution satisfies the second boundary condition in (10.5), then the solution of (10.6) with $s = \bar{s}$ will be the solution of our boundary value problem. $\bar{s}$ is usually found by selecting various numerical values for s, and solving the initial value problem (10.5) numerically with **ode45**, with an eye to finding a value for s for which the second boundary condition in (10.5) is satisfied. Toward this end, the following function M-file is useful:

```
function r = trial(s)
% finds y(1) given that y(0) = 0, y'(0) = s
rhs2 = @(t,y) [y(2); -t^2*y(2) - y(1)];
[ta ya] = ode45(rhs2, [0 1], [0 s]);
r = ya(end, 1);
```

We note that for certain simple examples (10.6) can be solved symbolically with **dsolve**. If the resulting solution is denoted by $y(t, s)$, then $\bar{s}$ can be found by solving $y(1, s) = 1$ with **fzero** or **solve**.

A second approach is to use **bvp4c**, a special differential equations solver for boundary value problems. **bvp4c** requires a "preprocessor" **bvpinit** to set an initial guess for the solution (in this case, just the constant vector **[1 0]**). The code that follows shows how to use **bvp4c** to find and plot the (numerical) solution, which is shown in Figure 10.3. Note that the boundary conditions are inserted in the form of a function **bcond** of **ya** and **yb**, the values of y $= [y; y']$ at the two boundary points. The solver starts with the initial guess and looks for a nearby solution of the differential equation for which **bcond** vanishes.

```
>> solinit = bvpinit(0:0.2:1,[1 0]);
>> bcond = @(ya, yb) [ya(1); yb(1)-1];
>> bvsol = bvp4c(rhs2, bcond, solinit);
>> tt = 0:0.025:1;
>> yy = deval(bvsol, tt);
>> plot(tt,yy(1,:))
```

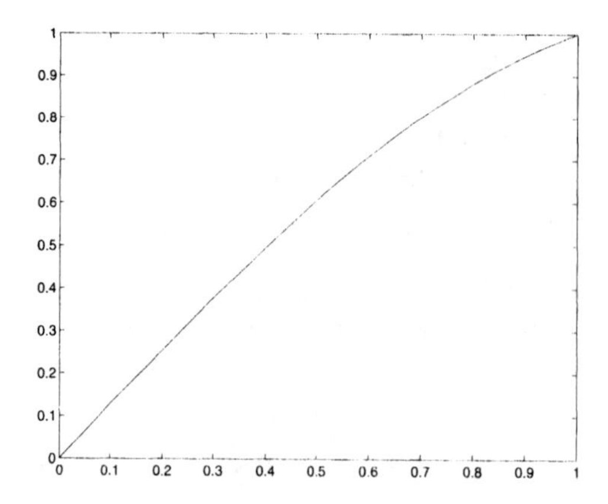

Figure 10.3: Solution of a Boundary Value Problem with **bvp4c**

10.2 Second Order Equations with Simulink

One can use Simulink, as well as **ode45**, to obtain numerical solutions of second order equations. We will illustrate how to do this with the linear initial value problem

$$\frac{d^2y}{dy^2} + (1+t^2)y = 0, \qquad y(0) = 0, \quad y'(0) = 1. \tag{10.7}$$

There are basically two ways to construct a model for this equation. The first, which yields the model shown in Figure 10.4, is the more obvious method. Namely, we use two

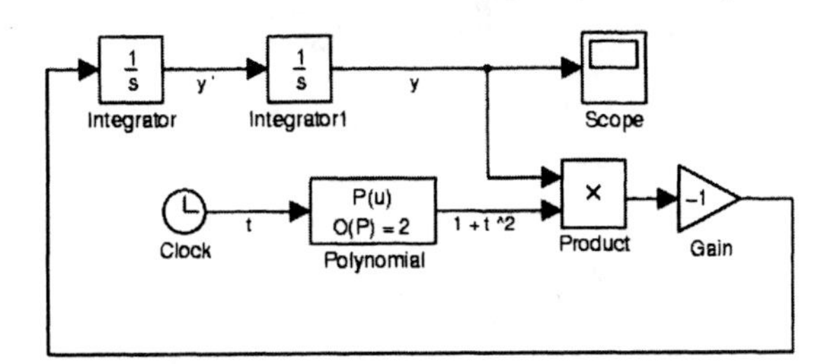

Figure 10.4: A Model for Equation (10.7)

Integrator blocks coupled together. The first Integrator has y'' as its input and y' as its output. The second has y' as its input and y as its output. Since the equation writes y'' as a function of y and t (or in most cases, also y', though the y' term is missing in (10.7)), that tells us how to link up the blocks. Here we have used a Polynomial Function block to compute $1 + t^2$ from t in a single step; then we multiply this by y in the Product block and then multiply by -1 in the Gain block to get y''. The initial condition in the first Integrator sets the value $y'(0) = 1$; the initial condition in the second Integrator (called Integrator1 here by default—Simulink always insists that all block names in a model be distinct) sets the value $y(0) = 0$. The Scope output from this model (with t running from 0 to 10) is shown in Figure 10.5. Note that the solutions are oscillatory; this is what we would expect from comparison with the constant-coefficient equation $y'' + y = 0$, whose solution (with the same initial conditions) is $y = \sin t$.

Another way to construct a Simulink model for a second order equation is to first convert the equation to a first order vector equation, which for (10.7) would be

$$(y, y')' = (y', -(1+t^2)).$$

If we substitute $\mathbf{y}$ for (y, y'), we can rewrite this using matrix operations as

$$\mathbf{y}' = \mathbf{y}\begin{pmatrix} 0 & -(1+t^2) \\ 1 & 0 \end{pmatrix}, \tag{10.8}$$

and we can further rewrite the coefficient matrix as

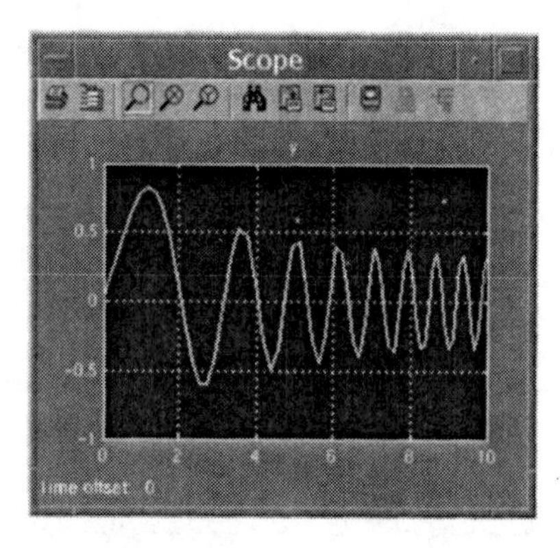

Figure 10.5: Scope Output with the Solution to (10.7)

$$\begin{pmatrix} 0 & -(1+t^2) \\ 1 & 0 \end{pmatrix} = (1+t^2)\begin{pmatrix} 0 & -1 \\ 0 & 0 \end{pmatrix} + \begin{pmatrix} 0 & 0 \\ 1 & 0 \end{pmatrix}. \tag{10.9}$$

This suggests constructing the model shown in Figure 10.6. This model uses vector signals

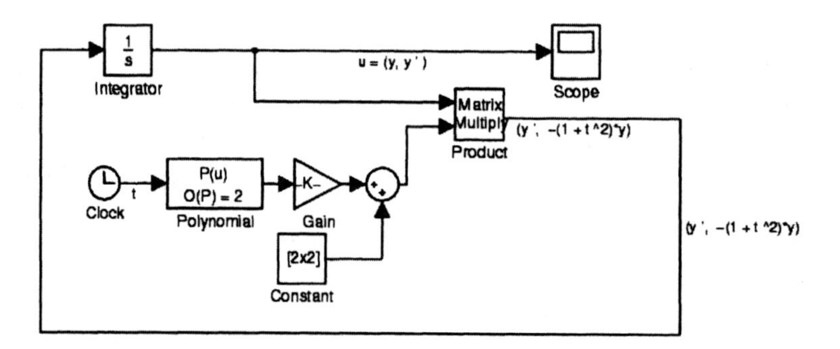

Figure 10.6: Vector Model for Equation (10.7)

and only a single Integrator. In this diagram, the Constant block has its block parameter set to `[0, 0; 1; 0]`, the Gain block has the gain set to `[0, -1; 0, 0]` and the multiplication type set to "Element-wise $K. * u$", in accordance with equation (10.9). Similarly, the Product block is set to matrix multiplication, in accordance with equation (10.8). Note that in the model of Figure 10.6, the Scope block is connected to a *vector signal*. The output one gets is shown in Figure 10.7. Note that one sees plots of both components of **y**, in other words of both y and y', as functions of time. Since y is the first component, it is plotted in yellow (the brighter curve in Figure 10.7); the second component y' appears in purple.

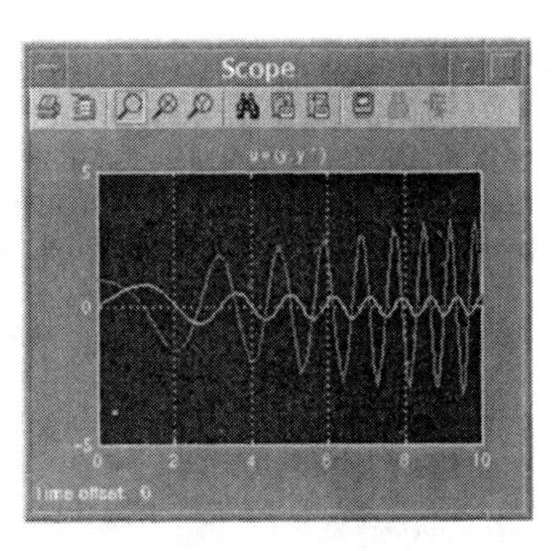

Figure 10.7: Scope Output with the Vector Solution to (10.7)

10.3 Comparison Methods

In this section, we discuss the Sturm Comparison Theorem. In addition, we will discuss the relation between the zeros of two linearly independent solutions of a second order linear equation.

Theorem 10.1 (Sturm Comparison Theorem) *Let $u(x)$ be a solution to the equation*

$$y'' + q(x)y = 0,$$

and suppose $u(a) = u(b) = 0$ for some $a < b$ (but u is not identically zero). Let $v(x)$ be a solution to the equation

$$y'' + r(x)y = 0,$$

where $r(x) \geq q(x)$ for all $a \leq x \leq b$. Then $v(x) = 0$ for some x in $[a, b]$.

We will defer the proof of this theorem briefly, preferring instead to explain how to use it. We typically use this result to compare a variable coefficient equation with an appropriate constant coefficient equation. Consider, for example, the equation

$$y'' + \frac{1}{x}y = 0. \tag{10.10}$$

Let K be a positive number. If $0 < x \leq K$, then $1/x \geq 1/K$. We now apply the above theorem with $q(x) = 1/K$ and $r(x) = 1/x$. Thus we compare equation (10.10) to the constant coefficient equation $y'' + (1/K)y = 0$, whose general solution is $u(x) = R\cos(\sqrt{1/K}x - \delta)$. Given any interval in $(0, K]$ of length $\pi\sqrt{K}$, the value of δ can be chosen so that $u(x) = 0$ at the endpoints of the interval. Then the Sturm Comparison Theorem implies that every solution of (10.10) has a zero on this interval. This argument implies that on $(0, K]$, the zeros of (10.10) cannot be farther apart than $\pi\sqrt{K}$.

In particular, solutions of (10.10) must oscillate as $x \to \infty$, though the oscillations may become less frequent as x increases. Indeed, by turning the above comparison around, one concludes that the zeros of solutions of (10.10) in $[K, \infty)$ must be at least $\pi\sqrt{K}$ apart.

Figure 10.8 shows two representative solutions of (10.10) computed with **dsolve**. As you can see, the graph confirms the predictions of the previous paragraph.

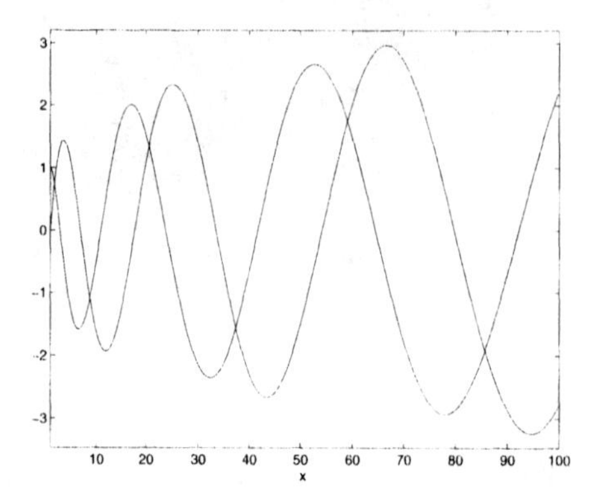

Figure 10.8: Solutions of $y'' + \frac{1}{x}y = 0$

The Sturm Comparison Theorem can also be used to study the zeros of solutions to Airy's equation and to Bessel's equation. In the latter case, a substitution is made to eliminate the y' term.

10.3.1 The Interlacing of Zeros

You may have noticed that the curves in Figure 10.8 take turns crossing the horizontal axis. We say that the zeros of the two solutions are *interlaced*: between any two zeros of one solution there is a zero of the other. This is a common phenomenon—the property of interlaced zeros holds for any pair of linearly independent solutions of a homogeneous second order linear differential equation

$$y'' + p(x)y' + q(x)y = 0. \tag{10.11}$$

If $p(x) = 0$, as is the case in equation (10.1), the proof of the interlacing result follows from an application of the Sturm Comparison Theorem by comparing the equation with itself. For the more general equation (10.11), the proof is based on the Wronskian. Let $y_1(x)$ and $y_2(x)$ be any two linearly independent solutions of (10.11). Recall that the Wronskian of $y_1(x)$ and $y_2(x)$ is

$$W(x) = y_1(x)y_2'(x) - y_1'(x)y_2(x).$$

Since y_1 and y_2 are solutions of (10.11), it follows that W satisfies the differential equation

$$W' = -p(x)W,$$

whose general solution is

$$W(x) = Ce^{-\int p(x)dx}.$$

This formula shows that $W(x)$ does not change sign—it always has the same sign as the constant C, which must be nonzero (or else y_1 and y_2 would be linearly dependent).

Assume that $y_1(a) = y_1(b) = 0$ and that there are no zeros of y_1 between a and b. Then

$$W(a) = -y_1'(a)y_2(a),$$
$$W(b) = -y_1'(b)y_2(b).$$

Since y_1 does not change sign between a and b, we know $y_1'(a)$ and $y_1'(b)$ have opposite signs. Then, since $W(a)$ and $W(b)$ have the same sign, $y_2(a)$ and $y_2(b)$ must have opposite signs. Therefore y_2 must be zero somewhere between a and b. Similarly, y_1 must have a zero between any two zeros of y_2.

Exercise 10.1 The solutions to Airy's equation and Bessel's equation are considered so important that they have been given names. The Airy functions $\mathrm{Ai}(x)$ and $\mathrm{Bi}(x)$ are a fundamental set of solutions for (10.1), and the Bessel functions $J_0(x)$ and $Y_0(x)$ are a fundamental set of solutions for (10.2). These functions are built into MATLAB as **`airy(0,x)`**, **`airy(2,x)`**, **`besselj(0,x)`**, and **`bessely(0,x)`**. Use MATLAB to check that the zeros of the two solutions **`airy(0,x)`** and **`airy(2,x)`** to Airy's equation are interlaced. Do the same for the Bessel functions **`besselj(0,x)`** and **`bessely(0,x)`**.

Remark 10.1 The Airy functions and the Bessel functions, along with the error function, and the Sine Integral function, are examples of *special functions*, which were introduced in Chapter 5.

10.3.2 Proof of the Sturm Comparison Theorem

The proof is based on the Wronskian, as was the proof of the interlacing result. Recall the hypotheses: $u'' + q(x)u = 0$ and $v'' + r(x)v = 0$, with $q(x) \leq r(x)$ between two zeros a and b of u. Since u is not identically zero, we can assume u does not change sign between a and b. (If it did it would have a zero between a and b, and we could look at a smaller interval.) Assume, for instance, that $u > 0$ between a and b. Since $u = 0$ at a and b, it follows that $u'(a) > 0$ and $u'(b) < 0$. (Notice that these derivatives cannot be zero because then by the uniqueness theorem u would be identically zero.) We want to prove that v is zero somewhere in $[a, b]$; we suppose it is not, say $v(x) > 0$ throughout $[a, b]$, and argue to obtain a contradiction.

Let $W(x)$ be the Wronskian of u and v,

$$W(x) = u(x)v'(x) - u'(x)v(x);$$

then

$$W(a) = -u'(a)v(a) < 0,$$
$$W(b) = -u'(b)v(b) > 0.$$

Also, since $q(x) \leq r(x)$,

$$W' = uv'' - u''v = u(-r(x)v) - (-q(x)u)v = (q(x) - r(x))uv \leq 0.$$

But this is impossible—the last inequality implies W does not increase between a and b, yet $W(a) < 0 < W(b)$. This contradiction means that v cannot have the same sign throughout $[a, b]$, and therefore v must be zero somewhere in the interval.

10.4 A Geometric Method

As mentioned above, direction fields are not directly applicable to second order equations. We now show, however, that with a given second order homogeneous equation, we can associate a first order equation whose direction field yields information about the solutions of the second order equation. (Our approach is similar to the classical method of associating a first order Riccati equation to a second order linear equation. The substitution we use is akin to the Prüfer substitution for Sturm-Liouville systems; see G. Birkhoff and G.-C. Rota, **Ordinary Differential Equations**, 3rd ed., J. Wiley and Sons, Inc., 1978.)

Consider the homogeneous equation

$$y'' + p(x)y' + q(x)y = 0. \tag{10.12}$$

Let

$$z = \arctan\left(\frac{y}{y'}\right);$$

then

$$z' = \left(1 + \left(\frac{y}{y'}\right)^2\right)^{-1} \frac{d}{dx}\left(\frac{y}{y'}\right) = \frac{y'^2}{y'^2 + y^2}\,\frac{y'^2 - yy''}{y'^2} = \frac{y'^2 - yy''}{y'^2 + y^2}.$$

Also, since $\tan z = y/y'$, notice that

$$\sin z = \frac{y}{\sqrt{y'^2 + y^2}}, \qquad \cos z = \frac{y'}{\sqrt{y'^2 + y^2}}.$$

Then if y satisfies (10.12) and y is not identically zero, it follows that

$$z' = \frac{y'^2 - yy''}{y'^2 + y^2} = \frac{y'^2 - y(-p(x)y' - q(x)y)}{y'^2 + y^2} = \frac{y'^2 + p(x)yy' + q(x)y^2}{y'^2 + y^2}.$$

In other words,

$$z' = \cos^2 z + p(x)\sin z\cos z + q(x)\sin^2 z. \tag{10.13}$$

We have shown that for every solution y of (10.12) that is not identically zero, there is a corresponding solution z of (10.13). (This is not a one-to-one correspondence—every constant multiple of a solution y corresponds to the same solution z.) Although the solution curves of (10.13) will be different from the solution curves of (10.12), we now show that we can learn about the solutions of (10.12) by studying the solutions of (10.13), specifically, by considering the direction field of (10.13).

10.4.1 The Constant Coefficient Case

We begin by considering the equation

$$y'' - y = 0,$$

whose general solution is

$$y = c_1 e^x + c_2 e^{-x}.$$

The corresponding first order equation is

$$z' = \cos^2 z - \sin^2 z.$$

In Figure 10.9, we show the direction field of this equation, which we produced with the MATLAB commands:

```
>> firstorder = @(x, z) ...
    cos(z).^2 - sin(z).^2;
>> [X, Z] = meshgrid(0:0.3:10, -pi/2:0.2:pi/2);
>> W = firstorder(X, Z);
>> L = sqrt(1 + W.^2);
>> quiver(X, Z, 1./L, W./L, 0.5)
>> axis tight
```

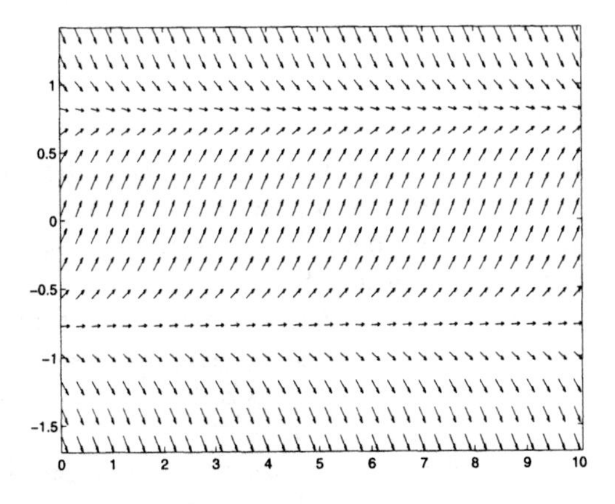

Figure 10.9: Direction Field of $z' = \cos^2 z - \sin^2 z$

Notice that we plot z from $-\pi/2$ to $\pi/2$; this is the range of values taken on by the arctan function. You may wonder what happens if z goes off the bottom of the graph. The answer is that z "wraps around" to the top of the graph. Recall that $z = \arctan(y/y')$; the points $z = \pm\pi/2$ correspond to $y' = 0$, and when y' changes sign, then so does z by passing from $-\pi/2$ to $\pi/2$ (or vice versa).

Next, observe that there is an unstable equilibrium at $z = -\pi/4$, which corresponds to $y/y' = \tan(-\pi/4) = -1$; this represents solutions of the form $y = ce^{-x}$. Similarly, the stable equilibrium $z = \pi/4$ corresponds to $y/y' = \tan(\pi/4) = 1$, which represents the solutions $y = ce^x$. The fact that most solutions of the equation satisfied by z (those with initial condition other than $-\pi/4$) approach the stable equilibrium $z = \pi/4$ corresponds to the fact that most solutions $y = c_1e^x + c_2e^{-x}$ (those with $c_1 \neq 0$) grow like e^x as x increases.

Another basic example is

$$y'' + y = 0,$$

for which the corresponding first order equation is

$$z' = \cos^2 z + \sin^2 z = 1.$$

That is, z simply increases linearly, although every time it reaches $\pi/2$ it wraps around to $-\pi/2$. This corresponds to the oscillation of solutions y; for every time $z = \arctan(y/y')$ passes through zero, then so must y, and vice versa.

Exercise 10.2 Consider the general second order linear homogeneous equation with constant coefficients:

$$ay'' + by' + c = 0.$$

Investigate how the roots of the characteristic equation, if real, correspond to equilibrium solutions for z. Show that if the roots of the characteristic equation are complex, then z' is always positive.

10.4.2 The Variable Coefficient Case

The examples above suggest ways to draw parallels between solutions z of (10.13) and solutions y of (10.12). First of all, if z is positive, then y and y' have the same sign. This implies that y is moving away from zero as x increases. We say in this case that y is *growing*, meaning that $|y|$ is increasing. Similarly, if z is negative then y and y' have opposite signs, which implies that y is moving toward zero as x increases—we say in this case that y is *decaying*, meaning that $|y|$ is decreasing. If z changes sign by passing through zero, then so does y, and if z passes through $\pm\pi/2$, then y' changes sign, showing that y has passed through a local maximum or minimum.

In light of these observations, let us summarize what we can predict about the *long-term behavior* of a solution y of (10.12) in terms of the corresponding solution z of (10.13).

- If z remains positive as x increases, then y grows away from zero.

- If z remains negative as x increases, then y decays toward zero.

- If z continues to increase from $-\pi/2$ to $\pi/2$ and wraps around to $-\pi/2$ again, then y *oscillates* as x increases.

Notice from (10.13) that $z' = 1$ whenever $z = 0$, so it is not possible for z to decrease through zero. Thus the long-term behavior of z will usually fall into one of the three categories above.

We can be more precise about the growth or decay rate of y in cases where z approaches a limiting value. If the limiting value is θ, then y/y' approaches $\tan\theta$ as x increases. In other words, $y' \approx (\cot\theta)y$ for large x. Thus

$$y \approx ce^{(\cot\theta)x}$$

for large x, and $\cot\theta$ is the asymptotic exponential growth (or decay) rate of y. If z approaches zero as x increases, then y/y' approaches zero as well. One can show that y grows (if $z > 0$) or decays (if $z < 0$) faster than any exponential function. Similarly, if z approaches $\pi/2$, then y grows slower than exponentially, and if z approaches $-\pi/2$, then y decays slower than exponentially. Since y'/y approaches zero in these cases, y could grow or decay toward a finite, nonzero value.

Let us now see what the direction field of (10.13) tells us about solutions of Airy's and Bessel's equations.

10.4.3 Airy's Equation

For Airy's equation,

$$y'' - xy = 0, \tag{10.14}$$

the corresponding first order equation is

$$z' = \cos^2 z - x\sin^2 z.$$

Figure 10.10 shows the direction field of this equation, plotted by a MATLAB command similar to the one used for Figure 10.9.

Notice that z is increasing steadily for negative x, so solutions of (10.14) must oscillate for negative x. For positive x, it is evident that solutions of (10.14) cannot oscillate; z passes through zero at most once, after which it remains positive. Hence the corresponding solution of (10.14) grows away from zero as x increases. It appears that once z becomes positive, it decreases to zero, indicating that most solutions of (10.14) grow faster than exponentially.

We now know there are solutions of (10.14) that grow very fast as x increases, and there are no oscillating solutions for positive x. Is it possible to have a decaying solution of (10.14)? Equivalently, is it possible for z to remain negative as x increases? Figure 10.10 strongly suggests that there is such a solution z, and hence that there is a solution y to (10.14) that decays toward zero at a rate faster than exponential.

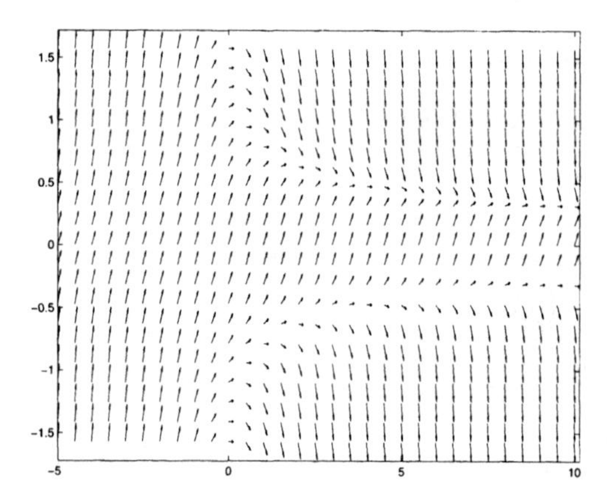

Figure 10.10: Direction Field of $z' = \cos^2 z - x \sin^2 z$

The Airy function $\mathrm{Ai}(x)$ is, by definition, the unique (up to a constant multiple) solution of (10.14) that decays as x increases. As expected, the other Airy function $\mathrm{Bi}(x)$ grows as x increases. You can see what these functions look like by plotting **`airy(x)`** and **`airy(2,x)`** in MATLAB. The general solution of (10.14) is a linear combination of $\mathrm{Ai}(x)$ and $\mathrm{Bi}(x)$.

10.4.4 Bessel's Equation

For Bessel's equation

$$y'' + \frac{1}{x}y' + y = 0, \tag{10.15}$$

the corresponding first order equation is

$$z' = \cos^2 z + \frac{1}{x} \sin z \cos z + \sin^2 z.$$

Figure 10.11 shows the direction field of this equation.

The picture is simpler than the one for Airy's equation. When x is away from zero, z is increasing steadily. So, solutions of (10.15) oscillate for both positive and negative x. Near $x = 0$, the direction field is irregular because of the $1/x$ term in the differential equation. We can't tell what happens from this picture, but there is reason to expect that solutions of (10.15) will not behave nicely near this singularity in the differential equation.

One question this approach cannot answer easily is whether the amplitude of oscillating solutions grows, decays, or remains steady as x increases. Since the coefficient of the y' term in (10.15) is positive, we can expect solutions of (10.15) to behave like a physical oscillator with damping—the amplitude of oscillations should decrease as x increases. Since the damping coefficient goes to zero as x increases, we cannot be sure (without a more refined analysis) whether the amplitude of solutions of (10.15) must decrease to zero.

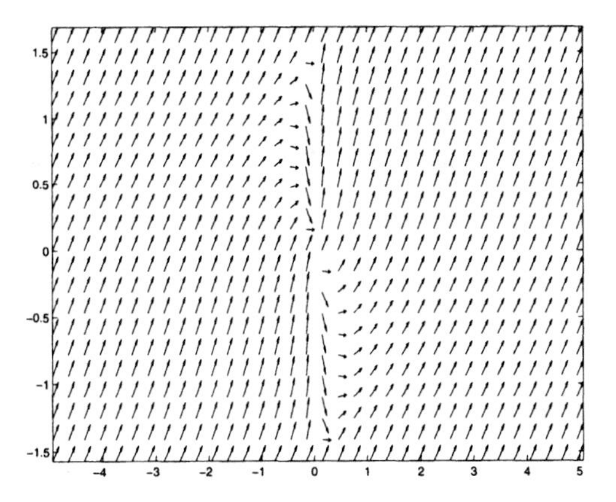

Figure 10.11: Direction Field of $z' = \cos^2 z + \frac{1}{x}\sin z \cos z + \sin^2 z$

In fact it does, as you can check by plotting the Bessel functions **besselj(0, x)** and **bessely(0, x)** in MATLAB.

10.4.5 Other Equations

Here are some other differential equations for which you can try this method.

- Bessel's equation of order n,

$$y'' + \frac{1}{x}y' + \left(1 - \frac{n^2}{x^2}\right)y = 0,$$

  for various n (above we studied the case $n = 0$). You can check your predictions using the functions **besselj(n,x)** and **bessely(n, x)**, which are built into MATLAB. These functions are considered in Problem 19 of Problem Set D.

- The modified Bessel equation of order n,

$$y'' + \frac{1}{x}y' - \left(1 + \frac{n^2}{x^2}\right)y = 0,$$

  for various n. The functions **besseli(n, x)** and **besselk(n, x)** are a fundamental set of solutions for this equation.

- The parabolic cylinder equation

$$y'' + \left(n + \frac{1}{2} - \frac{1}{4}x^2\right)y = 0$$

for various n. This equation arises in quantum mechanics, but its solutions are not built into MATLAB. They do exist in Maple and so MATLAB can solve the equation—but see Chapter 5 for the limitations on dealing with Maple functions that emerge from MATLAB commands. This equation is considered in Problem 17 of Problem Set D.

Problem Set D

Second Order Equations

The solution to Problem 1 appears in the *Sample Solutions* in the back of the book.

1. Airy's equation is the linear second order homogeneous equation $y'' = ty$. Although it arises in a number of applications, including quantum mechanics, optics, and waves, it cannot be solved exactly by the standard symbolic methods. In order to analyze the solution curves, let us reason as follows.

   (a) For t close to zero, the equation resembles $y'' = 0$, which has general solution $y = c_1 t + c_2$. We refer to this as a "facsimile" solution. Graph a numerical solution to Airy's equation with initial conditions $y(0) = 0$, $y'(0) = 1$, and the facsimile solution (with the same initial data) on the interval $(-2, 2)$. How well do they match?

   (b) For $t \approx -K^2 \ll 0$, the equation resembles $y'' = -K^2 y$, and the corresponding facsimile solution is given by $y = c_1 \sin(Kt + c_2)$. Again using the initial conditions $y(0) = 0$, $y'(0) = 1$, plot a numerical solution of Airy's equation over the interval $(-18, -14)$. Using the value $K = 4$, try to find values of c_1 and c_2 so that the facsimile solution matches well with the actual solution. Why shouldn't we expect the initial conditions for Airy's equation to be the appropriate initial conditions for the facsimile solution?

   (c) For $t \approx K^2 \gg 0$, Airy's equation resembles $y'' = K^2 y$, which has solution $y = c_1 \sinh(Kt + c_2)$. (The hyperbolic sine function is called **`sinh`** in MATLAB.) Plot a numerical solution of Airy's equation together with a facsimile solution (with $K = 4$) on the interval $(14, 18)$. In analogy with part (b), you have to choose values for c_1 and c_2 in the facsimile solution.

   (d) Plot the numerical solution of Airy's equation on the interval $(-20, 2)$. What does the graph suggest about the frequency and amplitude of oscillations as $x \to -\infty$? Could any of that information have been predicted from the facsimile analysis?

2. Consider Bessel's equation of order zero

$$t^2y'' + ty' + t^2y = 0 \tag{D.1}$$

with initial data $y(0) = 1,\ y'(0) = 0$. The solution is the Bessel function of order zero of the first kind, $J_0(t)$. In this problem we solve equation (D.1) and two approximations of it with **dsolve** to learn about $J_0(t)$. Strictly speaking, this equation has a singularity at $t = 0$. However, this is one instance of a solution to a linear equation that exists outside the expected domain of definition. The singularity causes no difficulty in this problem.

   (a) For t close to 0, t^2 is very small compared to t. The equation (D.1) is therefore approximately $ty' = 0$. Solve this equation with the preceding initial data. What does this "facsimile" solution to the original problem suggest to you about the behavior of J_0 near near $t = 0$?

   (b) For t large and positive, t is small compared to t^2; and so we may approximate (D.1) by the equation $t^2(y'' + y) = 0$. Solve this equation with the same initial data. What does this suggest about the nature of the function J_0 for large t?

   (c) Still thinking of t as large and positive, rewrite (D.1) in the form

$$y'' + \frac{1}{t}y' + y = 0.$$

   If $t \approx K \gg 0$, we might approximate the equation by the constant coefficient equation

$$y'' + \frac{1}{K}y' + y = 0.$$

   Choose a specific value for K, say $K = 100$, and solve this equation and see what further information you obtain about $J_0(t)$ for large positive t.

   (d) Solve Bessel's equation with the given initial conditions, and plot the solution. Are the conclusions of your analysis confirmed?

3. This and some of the following problems concern models for the motion of a pendulum, which consists of a weight attached to a rigid arm of length L that is free to pivot in a complete circle. Neglecting friction and air resistance, the angle $\theta(t)$ that the arm makes with the vertical direction satisfies the differential equation

$$\theta''(t) + \frac{g}{L}\sin(\theta(t)) = 0, \tag{D.2}$$

where $g = 32.2$ ft/sec^2 is the gravitational acceleration constant. We will assume the arm has length 32.2 ft and so replace (D.2) by the simpler form

$$\theta'' + \sin\theta = 0. \tag{D.3}$$

(Alternatively, one can rescale time, replacing t by $\sqrt{g/L}\,t$, to convert (D.2) to (D.3).) For motions with small displacements (θ small), $\sin\theta \approx \theta$, and (D.3) can be approximated by the linear equation

$$\theta'' + \theta = 0. \tag{D.4}$$

This equation has general solution $\theta(t) = A\cos(t-\delta)$, with *amplitude* A and *phase shift* δ. Hence all the solutions to the linear approximation (D.4) have period 2π, independent of the amplitude A. In this problem we consider solutions of equation (D.3) satisfying the initial conditions $\theta(0) = A$, $\theta'(0) = 0$. If $|A| < \pi$, these solutions are periodic. However, in contract to the linear equation (D.4), their periods depend on the amplitude A. We do expect that, for small displacements A, the solutions to the pendulum (D.3) will have periods close to 2π.

(a) Investigate how the period depends on the amplitude A by plotting a numerical solution of equation (D.3) using initial conditions $\theta(0) = A$, $\theta'(0) = 0$ on an appropriate interval for various A. Estimate the periods of the pendulum for the amplitudes $A = 0.1, 0.7, 1.5$, and 3.0. Confirm these results by displaying the displacements at a sequence of times, and finding the time at which the pendulum returns to its original position.

(b) The period is given by the formula

$$T = 4\int_0^{\pi/2} \frac{d\phi}{\sqrt{1-k^2\sin^2\phi}},$$

where $k = \sin(A/2)$. This formula may be derived in your text; it can be found in Section 9.3 of Boyce & DiPrima, Problem 27. The integral is called an *elliptic integral*. It cannot be evaluated by an elementary formula, but it can be evaluated numerically using **quad** or **quadl**. Calculate the period for the values of A we are considering. Do the values agree with those obtained in part (a)?

(c) Redo the numerical calculations in part (a) with different tolerances, choosing the tolerances so that the values you get agree with those calculated in part (b).

(d) How does the period depend on the amplitude of the initial displacement? For A small, is the period close to 2π? What is happening to the accuracy of the linear approximation as the initial displacement increases?

4. In this problem, we'll look at what the pendulum does for various initial velocities (*cf.* Problem 3).

(a) Numerically solve the differential equation (D.3) using initial conditions $\theta(0) = 0$, $\theta'(0) = 1$. Solve equation (D.4) with the same initial conditions. Plot on the same graph the solutions to both the nonlinear equation (D.3) and the linear equation (D.4) on the interval from $t = 0$ to $t = 40$, and compare the two. Be clear about which curve is the nonlinear solution and which is the linear solution.

(b) Repeat part (a) and compare the linear and nonlinear solutions for each of the following values of the initial velocity v: 1, 1.99, 2, 2.01. For the (numerical) nonlinear solution, interpret what the graph indicates the pendulum is doing physically. What do you think the exact solution does in each case?

5. In this problem, we'll investigate the effect of damping on the pendulum, using the model

$$\theta'' + b\theta' + \sin\theta = 0.$$

Prepare Simulink models that will take the value of b from a "Constant" block and

- plot the numerical solution of this differential equation with initial conditions $\theta(0) = 0$, $\theta'(0) = 4$, from $t = 0$ to $t = 20$;
- do the same for the linear approximation

$$\theta'' + b\theta' + \theta = 0.$$

(You do not need to send the plot to the printer if it displays in a Scope.) Compare the linear and nonlinear behavior for the values $b = 1$, 1.5, and 2. Interpret what is happening *physically* in each case; *i.e.*, describe explicitly what the graph says the pendulum is doing.

6. In this problem, we'll look at the effect of a periodic external force on the pendulum, using the model

$$\theta'' + 0.05\theta' + \sin\theta = 0.3\cos\omega t \tag{D.5}$$

(*cf.* Problems 3–5). We have chosen a value for the damping coefficient that is more typical of air resistance than the values in the previous problem. Prepare Simulink models for (D.5) and its linear approximation

$$\theta'' + 0.05\theta' + \theta = 0.3\cos\omega t.$$

The right-hand side can be produced by a Sine Wave block (which is in the Sources Library). When you install this block and left-click on it to bring up the **Block Parameters** menu, you will see an Amplitude box in which to insert the parameter 0.3 and a Frequency box in which to insert the parameter ω. Note that since we have a cosine, not a sine, you also have to adjust the Phase (to $\frac{\pi}{2}$, since $\cos\omega t = \sin(\omega t + \frac{\pi}{2})$).

- Plot the numerical solution of this differential equation with initial conditions $\theta(0) = 0$, $\theta'(0) = 0$, from $t = 0$ to $t = 60$.
- Do the same for the linear approximation.

(You can send the plots to the printer from a Scope display using the "printer" icon in the upper-left corner.)

Compare the nonlinear and linear models for the following values of the frequency ω: $0.6, 0.8, 1, 1.2$. Which frequency moves the pendulum farthest away from its equilibrium position? For which frequencies do the linear and nonlinear equations have widely different behaviors? Which forcing frequency seems to induce resonance-type behavior in the pendulum? Graph that solution on a longer interval and decide whether the amplitude goes to infinity.

7. In this problem, we study the effects of air resistance.

   (a) A paratrooper steps out of an airplane at a height of 1000 ft and after 5 seconds opens her parachute. Her weight, with equipment, is 195 lbs. Let $y(t)$ denote her height above the ground after t seconds. Assume that the force due to air resistance is $0.005y'(t)^2$ lbs in free fall and $0.6y'(t)^2$ lbs with the chute open. At what height does the chute open? How long does it take to reach the ground? At what velocity does she hit the ground? (This model assumes that air resistance is proportional to the *square* of the velocity and that the parachute opens instantaneously.)

   (*Hint*: This problem can be solved most efficiently by using an ODE file to detect significant events. Pay attention to units. Recall that the mass of the paratrooper is $195/32$, measured in lb sec^2/ft. Here, 32 is the acceleration due to gravity, measured in ft/sec^2.)

   (b) Let $v = y'$ be the velocity during the second phase of the fall (while the chute is open). One can view the equation of motion as an autonomous first order ODE in the velocity:

   $$v' = -32 + \frac{192}{1950}v^2.$$

   Make a qualitative analysis of this first order equation, finding in particular the critical or equilibrium velocity. This velocity is called the terminal velocity. How does the terminal velocity compare with the velocity at the time the chute opens and with the velocity at impact?

   (c) Assume the paratrooper is safe if she strikes the ground at a velocity within 5% of the terminal velocity in (b). Except for the initial height, use the parameters in (a). What is the lowest height from which she may parachute safely? (*Please do not try this at home!*)

8. In many applications, second order equations come with *boundary conditions* rather than initial conditions. For example, consider a cable that is attached at each end to a post, with both ends at the same height (let us call this height $y = 0$). If the posts are located at $x = 0$ and $x = 1$, the height y of the cable as a function of x satisfies the differential equation

$$y'' = c\sqrt{1 + (y')^2},$$

with boundary conditions $y(0) = 0$, $y(1) = 0$. The constant c depends on the length of the cable; for this problem we'll use $c = 1$.

(a) Solve the boundary value problem with **dsolve**.

(b) Solve this problem numerically with the shooting method. Plot the solution on $[0, 1]$, and determine the maximum dip in the cable.

(c) Solve the auxiliary initial value problem using **dsolve**, and then find the critical value for the parameter s using **fzero**.

(d) Solve this boundary value problem using **bvp4c**.

Show that the solutions obtained in parts (a), (b), (c), and (d) agree with each other.

9. As explained in the last problem, sometimes one is faced with a second order differential equation with a boundary condition. Existence and uniqueness of solutions for such problems is more complicated than for initial value problems, as you will see in this problem. Consider the simple boundary value problem (BVP)

$$y'' + \alpha^2 y = 0, \qquad y(0) = 0, \quad y(1) = 1, \tag{D.6}$$

where $\alpha > 0$ is a parameter.

(a) Solve the problem explicitly with **dsolve** using only the left-hand condition $y(0) = 0$. Since this is a second order equation and you have not specified $y'(0)$, the answer should involve an undetermined constant $C1$. Solve for this constant (in terms of α) using **solve** in order to satisfy the right-hand condition $y(1) = 1$. For what values of α does a solution exist?

(b) Redo part (a), but now modifying the right-hand condition in (D.6) to $y(1) = 0$. This time there is always a solution (namely $y = 0$), since the equation is now linear and homogeneous, but sometimes it is non-unique. For what values of α do you get more than one solution of the BVP? How much non-uniqueness is there in the solution?

(c) What happens to the solution of (D.6) (as computed in part (a)) when $\alpha \to \pi$? What do you observe about the solution as computed with **bvp4c** for α close to π, say 3.1415926? What about $\alpha = 3.14159265358898$? How do the plots of the solutions in the two cases differ, and how can you explain this?

10. The problem of finding the function $u(x)$ satisfying

$$\begin{cases} a(x)u''(x) + a'(x)u'(x) = f(x), & 0 \le x \le 1 \\ u(0) = u(1) = 0. \end{cases} \tag{D.7}$$

arises in studying the longitudinal displacements in a longitudinally loaded elastic bar. The bar is of length 1, its left end is at $x = 0$, and its right end at $x = 1$. In the differential equation above, $f(x)$ represents the external force on the bar (which is assumed to be longitudinal, *i.e.*, directed along the bar), $a(x)$ represents both the elastic properties and the cross-sectional area of the bar, and $u(x)$ is the longitudinal displacement of the bar at the point x. This problem is an example of a *boundary value problem*. The function $a(x)$ may be constant; this is the case if neither the elastic properties nor the cross-sectional area depends on the position x in the rod, *i.e.*, if the rod is uniform. But if the rod is not uniform, then $a(x)$ is not constant, and the equation has variable coefficients. Similarly, $f(x)$ will be constant only if the external force is applied uniformly along the rod.

(a) Take $a(x) = 1 + x$ and $f(x) = 5\sin^2(2\pi x)$. Here the force is applied symmetrically and is strongest at $x = 0.25$ and $x = 0.75$, as if someone is hanging by both hands from a relatively short rod. Use **dsolve** to find a solution $u(x)$ to the boundary problem (D.7), and plot it on the interval $[0, 1]$. Based on the result, do you think the rod is more flexible where $a(x)$ is larger or where it is smaller?

(b) Now suppose $a(x) = 1 + \exp(x)$ and $f(x)$ is the same as in part (a). Find and plot the corresponding solution $u(x)$ to (D.7). Since MATLAB cannot solve this equation symbolically, we will use a numerical method. We cannot apply **ode45** directly, since it requires initial conditions. Nonetheless, we can use the shooting method to find the value of $u'(0)$ that leads to a solution satisfying the condition $u(1) = 0$. By trial and error, find $u'(0)$ to at least two decimal places, and graph the resulting solution on $[0, 1]$. What is the maximum displacement, and where does it occur?

(c) Repeat part (b) using **bvp4c** instead of the shooting method.

11. This problem is based on Problem 32 in Boyce & DiPrima, Section 3.8. Consider a frictionless mass-spring system as in Figure 3.8.10 in Boyce & DiPrima (standard mass attached to a spring on a frictionless table). Suppose the restoring force of the spring is not given by Hooke's Law, but instead is of the form

$$F = -(ky + \epsilon y^3).$$

If $\epsilon = 0$, the assumption amounts to Hooke's Law, but in this problem we shall focus on $\epsilon \neq 0$ (either positive or negative). If we have air resistance present with damping coefficient γ, then the equation of motion (assuming the displacement is indicated by the variable y) becomes

$$my'' + \gamma y' + ky + \epsilon y^3 = 0$$

(see Boyce & DiPrima, 8th edition, p. 206). Henceforth we shall normalize by assuming that $m = k = 1$ and $\gamma = 0$, and then take the initial conditions

$$y(0) = 0, \quad y'(0) = 0.$$

(a) Plot the solution when $\epsilon = 0$. What is the amplitude and period of the solution?

(b) Let $\epsilon = 0.1$. Plot a numerical solution. Is the motion periodic? Estimate the amplitude and period.

(c) Repeat part (b) for $\epsilon = 0.2$, and then for $\epsilon = 0.3$.

(d) Plot your estimated values of the amplitude A and period T as functions of ϵ. How do A and T depend on ϵ?

(e) Repeat parts (b)–(d) for negative values of ϵ.

12. In this problem we study how solutions of the initial value problem

$$\frac{d^2y}{dt^2} + 0.15\frac{dy}{dt} - y + y^3 = 0, \quad y(0) = c, \quad y'(0) = 0$$

depend on the initial value c.

(a) Plot a numerical solution of this equation from $t = 0$ to $t = 40$ for each of the initial values $c = 0.5, 1, 1.5, 2, 2.5$. Describe how increasing the initial value of y affects the solutions, both in terms of their limiting behavior and their general appearance.

(b) Plot all five solutions on one graph. Would such a picture be possible for solutions of a first order differential equation? Why or why not?

13. In this problem, we consider the long-term behavior of solutions of the initial value problem

$$\frac{d^2y}{dt^2} + 0.2\frac{dy}{dt} - y + y^3 = 0.3\cos(\omega t), \quad y(0) = 0, \quad y'(0) = 0 \tag{D.8}$$

for various frequencies ω in the forcing term.

Plot a numerical solution of (D.8) from $t = 0$ to $t = 100$ for each of the eight frequencies $\omega = 0.8, 0.9, \ldots, 1.4, 1.5$.

Describe and compare the different long-term behaviors you see. Due to the forcing term, all solutions will oscillate, but pay particular attention to the magnitude of the oscillations and to whether or not there is a periodic pattern to them. Are there any similarities between your results for this nonlinear system and the phenomenon of resonance for linear systems with periodic forcing?

14. In this problem, we study the zeros of solutions of the second order differential equation

$$y'' + (3 - \cos t)y = 0. \tag{D.9}$$

(a) Compute and plot several solutions of this equation with different initial conditions: $y(0) = c$, $y'(0) = d$. To be specific, choose three different values for the pair c, d and plot the corresponding solutions over $[0, 20]$. By inspecting your plots, find a number L that is an upper bound for the distance between successive zeros of the solutions. Then find a number l that is a lower bound for the distance between successive zeros.

(b) Information on the zeros of solutions of linear second order ODEs can be obtained from the Sturm Comparison Theorem (see Theorem 10.1 in Chapter 10). By comparing (D.9) with the equation $y'' + 2y = 0$, you should be able to get a value for L, and by comparing (D.9) with $y'' + 4y = 0$, you should be able to find l. How do these values compare with the values obtained in part (a)? (You will find it useful to note that the general solution of $y'' + ky = 0$, where k is a positive constant, can be written as $y = R\cos(\sqrt{k}x - \delta)$, with $R \geq 0$, δ arbitrary. R is called the *amplitude* and δ the *phase shift*.)

(c) Plot a solution of (D.9) and a solution of $y'' + 2y = 0$ on the same graph, and verify that between any two zeros of the latter solution, there is at least one zero of the solution of (D.9).

15. This problem is based on Problems 18 and 19 in Boyce & DiPrima, Section 3.9. Consider the initial value problem

$$u'' + u = 3\cos(\omega t), \quad u(0) = 0, \quad u'(0) = 0.$$

(a) Find the solution (using **dsolve**). For $\omega = 0.5, 0.6, 0.7, 0.8, 0.9$, plot the solution curves on the interval $0 \leq t \leq 15$. Note that $\omega_0 = 1$ is the natural frequency of the homogeneous equation. Describe how the solution curves change as ω gets closer to 1.

(b) Note that the formula you found in part (a) is invalid when $\omega = 1$. Find and plot the solution curve for $\omega = 1$ on the interval $0 \leq t \leq 15$. Based on the discussion of forced vibrations in your text, what phenomenon should be exhibited for this value of ω? Corroborate your answer by plotting on a longer interval.

(c) Plot the solution for $\omega = 0.9$ on a longer interval and compare it with the solution from part (b). What phenomenon is exhibited by the curve for $\omega = 0.9$?

16. A solution of a second order linear differential equation is called *oscillatory* if it changes sign infinitely many times and *nonoscillatory* if it changes sign only finitely many times. In this problem, we will be interested in determining the oscillatory nature of nonzero solutions to some second order linear ODEs. First consider the equation $y'' + ky = 0$, where k is a constant. If $k > 0$, it has the general solution $y = R\cos(\sqrt{k}x - \delta)$, with $R \geq 0, \delta$ arbitrary. The constant R is called the *amplitude*, and δ is called the *phase shift*. From this formula we see that every solution $y(x)$ has infinitely many zeros, and hence changes sign infinitely many times, *i.e.*, is oscillatory. If $k < 0$, the general solution is $y(x) = c_1 e^{\sqrt{-k}x} + c_2 e^{-\sqrt{-k}x}$, while if $k = 0$, the general solution is $y(x) = c_1 x + c_2$. These solutions change sign at most once, and so are nonoscillatory.

Next, consider Airy's equation

$$y'' = xy, \tag{D.10}$$

which arises in various applications. Since (D.10) has a variable coefficient, we cannot study it by elementary methods; in particular, we cannot find the general solution in terms of elementary functions as we did for the constant coefficient equation. We will instead first make a graphical study of the solutions of (D.10) and then study the solutions using the Sturm Comparison Theorem (see Chapter 10).

(a) Compute and plot several solutions of (D.10) with different initial conditions $y(0) = c$, $y'(0) = d$ over the interval $[-10, 5]$. Use different combinations of positive, negative, and zero values for c and d. What can you say in general about the zeros of the solution? Where do they occur on the x axis, and how far apart are the successive zeros of a solution? Are the solutions oscillatory as $x \to \infty$? As $x \to -\infty$?

(b) Information on the zeros of solutions of linear second order ODEs, and hence information on their oscillatory nature, can also be obtained from the Sturm Comparison Theorem. By comparing (D.10) with the equation $y'' + by = 0$ for $x \leq -b < 0$, what do you learn about zeros on the negative x axis and their spacing, and hence about the oscillatory nature of solutions of (D.10) for $x \leq 0$? By comparing (D.10) with $y'' = 0$ for $0 < x$, what do you learn about the oscillatory nature of solutions for $x > 0$? How many zeros could a solution have on the positive x-axis? Do the graphs you plotted in part (a) agree with these results?

17. In this problem, we study solutions of the parabolic cylinder equation

$$y'' + \left(n + \frac{1}{2} - \frac{x^2}{4}\right) y = 0, \tag{D.11}$$

which arises in the study of quantum-mechanical vibrations. Since the equation (D.11) is unchanged if x is replaced by $-x$, any solution function will be symmetric with respect to the y-axis. Therefore, we focus our attention on $x \geq 0$.

(a) Find the corresponding first order equation for $z = \arctan(y/y')$, as described in Section 10.4.

(b) For $n = 1$, plot the direction field for the z equation from $x = 0$ to $x = 10$. (Remember to use $-\pi/2 \leq z \leq \pi/2$.) Based on the plot, predict what the solutions y to the parabolic cylinder equation look like near $x = 0$ and for larger x. Is there a value of x around which you expect their behavior to change?

(c) Now solve the parabolic cylinder equation numerically for $n = 1$ with the two sets of initial conditions $y(0) = 1$, $y'(0) = 0$ and $y(0) = 0$, $y'(0) = 1$. Plot the two solutions on the same graph. (It will probably help to change the range on the plot.) Do the solutions behave as you expected?

(d) Repeat parts (b) and (c) for $n = 5$. Point out any differences from the case $n = 1$.

(e) Repeat parts (b) and (c) for $n = 15$. Discuss how the solutions are changing as n increases.

(f) Now consider the three solutions (for $n = 1$, 5, and 15) with $y(0) = 0$. Point out similarities and differences. By drawing an analogy with Airy's equation, argue from the direction fields that, for any n, exactly one solution function decays for large x while all others grow. Do you have enough graphical evidence from the numerical plots to conclude that the solution corresponding to the initial data $y(0) = 0$, $y'(0) = 1$ is that function? Why or why not? (*Hint*: Look at previous discussions of stability of solutions as a guide.)

18. Consider Bessel's equation of order zero

$$x^2y'' + xy' + x^2y = 0.$$

(a) Compute and plot several solutions of this equation with different initial conditions: $y(0.1) = c, y'(0.1) = d$. To be specific, choose three different values for the pair c, d and plot the corresponding solutions over $[0.1, 20]$. (Why isn't it a good idea to use $x = 0$ in the initial conditions?) By inspecting your plots, find a number L that is an upper bound for the distance between successive zeros of the solutions. Then find a number l that is a lower bound for the distance between successive zeros.

(b) Now confirm your findings with the Sturm Comparison Theorem. It doesn't apply directly, but if we introduce the new function $z(x) = x^{1/2}y(x)$, Bessel's equation becomes

$$z'' + (1 + 1/(4x^2))z = 0,$$

which has the form of the equation in the Comparison Theorem. Since y will have a zero wherever z has a zero, we can study the zeros of y by studying the zeros of z. By comparing the equation for z with the equation $z'' + z = 0$, determine an upper bound L on the distance between successive zeros of any solution of Bessel's equation.

(c) Let a be a positive number. For $x \in [a, \infty)$, the quantity $1+1/(4x^2)$ is less than or equal to the constant $1 + 1/(4a^2)$. By making an appropriate comparison, determine a lower bound l on the distance between successive zeros of any solution of Bessel's equation (for $x > a$). What is the limiting value of l as a goes to ∞? Approximately how far apart are the zeros when x is large? Did your graphical study lead to comparable values for l and L?

19. In this problem, we study solutions of Bessel's equation of order n,

$$y'' + \frac{1}{x}y' + \left(1 - \frac{n^2}{x^2}\right)y = 0, \qquad \text{(D.12)}$$

for $n > 0$. Solutions of this equation, called *Bessel functions* of order n, are used in the study of vibrations and waves with circular symmetry. Since (D.12) is unchanged if x is replaced by $-x$, we focus our attention on $x \geq 0$.

(a) Find the corresponding first order equation for $z = \arctan(y/y')$, as described in Section 10.4 of Chapter 10.

(b) For $n = 1$, plot the direction field for the z equation from $x = 0$ to $x = 20$. (Remember to use $-\pi/2 \le z \le \pi/2$.) Based on the plot, predict what the Bessel functions of order 1 look like for small x and then for large x. Is there a value of x around which you expect their behavior to change?

(c) Now plot the Bessel functions **besselj(n, x)** and **bessely(n, x)** for $n = 1$ on the same graph. (It will probably help to change the range on the plot.) Do the solutions behave as you expected?

(d) Repeat parts (b) and (c) for $n = 5$. Point out any differences from the case $n = 1$.

(e) Repeat parts (b) and (c) for $n = 15$. Discuss how the solutions are changing as n increases.

20. Consider the mass-spring system with dry friction, depicted in Figure D.1. This

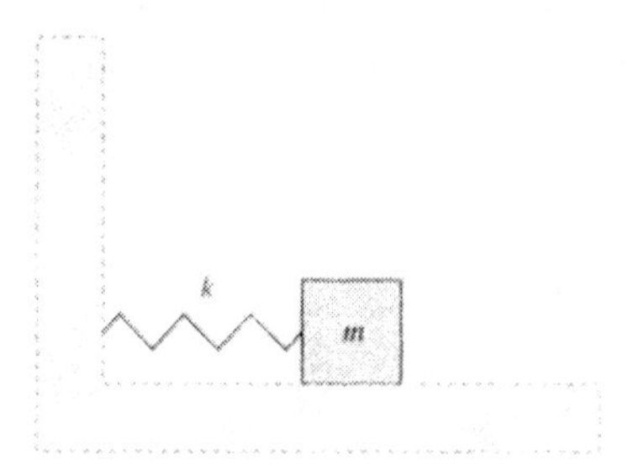

Figure D.1: Mass-Spring System with Friction

system is governed by the following equations:

$$\begin{cases} my'' = -ky - \mu mg\,\mathrm{sign}(y'), & \text{unless } y' = 0 \text{ and } |ky| < \mu mg, \\ y'(t) \equiv 0, \quad t \ge t_0 & \text{if } y'(t_0) = 0 \text{ and } |ky(t_0)| < \mu mg. \end{cases} \tag{D.13}$$

The first equation in (D.13) governs *kinetic friction*, and says that the object of mass m is subject to a restoring force opposite and proportional to the displacement y, with proportionality constant the spring constant k, as well as to a friction force opposite to the direction of motion, given by the friction coefficient μ times the force of gravity on the mass. The second equation in (D.13) governs *static friction*, and says that if the mass is instantaneously not moving at time t_0 and the force exerted on it by the spring is not big enough, then it will never move at all. In this model we are assuming for simplicity (even though this is usually not the case) that the coefficients of static and kinetic friction are equal. (In a more realistic model, the μ in the second equation would be larger than the μ in the first equation.) We are also assuming that the force of friction is independent of velocity in the kinetic friction case, which is not 100% correct, but is a reasonable first approximation to the truth. Note that the equations (D.13) really only depend on two constants, $d = \mu g$, and $\omega = \sqrt{k/m}$. When $\mu = 0$ (case of no friction), they reduce to the usual spring equation $y'' + \omega^2 y = 0$ with angular frequency ω. On the other hand, in the limiting case where $\mu \to \infty$, the object comes to an immediate stop and then doesn't move at all.

(a) Write a function M-file `friction.m` starting with the line

```
function [T, Y] = friction(time, d, omega, y0, v0)
```

that takes as parameters the time interval **time**, d, ω, and initial values **y0** of position and **v0** of velocity, and outputs a table of values of y with the corresponding values of position and velocity. There are (at least) two ways to structure your M-file. The most natural method is to have lines that say

```
options = odeset('Events', static);
[T, Y, TE, YE, IE] = ...
    ode45(rhs, time, [y0 v0], options);
```

Ignoring the **options** for the moment, the second line calls **ode45** on an anonymous function **rhs** (whose definition will involve the parameters **d** and **omega**) that represents the right-hand side of the kinetic friction equation, expressed as a first order system. But **odeset** is used here to call the event detection feature of the ODE solver, to check for the condition

$$|y| < \frac{\mu m g}{k} = \frac{d}{\omega^2}$$

in the static friction equation. This condition should be encoded in an anonymous function **static**. Since **ode45** insists that an "events" function have only two inputs, and since the static friction condition also involves **d** and **omega**, the easiest thing is to define **static** by a line

```
static = @(t, y) static1(t, y, d, omega)
```

and to define an additional function M-file `static1.m` just as in Section 8.6.3, but with four inputs instead of two. The first output of `static1.m` should be **y(2)**, the velocity, but the second output, the "stop function," should be 1 if **abs(y(1))** is small enough (meaning one should stop in this case), 0 otherwise. Once the integration of the kinetic friction equation stops, the mass should remain constant thereafter. The alternative method is to call **ode45** on a "right-hand side" function that is discontinuous, and is given by the two different equations in (D.13), depending on conditions.

(b) Suppose we denote the initial data by $y_0 = y(0)$, $v_0 = y'(0)$. Use your function M-file `friction.m` to solve the IVP when $y_0 = 4$, $v_0 = 0$, $\omega = 1$, and $d = 0.75$. Draw two graphs: on one, graph the displacement y for $0 \le t \le 10$; on the second, draw a phase diagram for y, y' over the same time interval. Explain what the pictures mean. Where does the mass come to rest?

(c) Answer part (b) but with $y_0 = 3$, then with $y_0 = 2$.

(d) Now change the initial data to $y_0 = 0,\ v_0 = 5$ and leave $\omega = 1, d = 0.75$. What kind of motion ensues? (*Hint*: You may need to lengthen the time interval.)

(e) Find (to the nearest tenth) the value of v_0 (with $y_0 = 0, \omega = 1, d = 0.75$) such that initial velocities below or equal to that value do not propel the mass out of the "friction well" $-0.75 \leq y \leq 0.75$, but initial velocities above it propel it so that that it goes beyond the well once, but when it returns to the well, it does not escape it again.

Chapter 11

Series Solutions

A primary theme of this book is that numerical, geometric, and qualitative methods can be used to study solutions of differential equations, even when we cannot find an exact formula solution. Of course, formula solutions are extremely valuable, and there are many techniques for finding them. For instance, techniques for finding formula solutions of second order linear differential equations include:

- The exponential substitution, which leads to solutions of an arbitrary constant coefficient homogeneous equation.
- The method of *reduction of order*, which produces a second linearly independent solution of a homogeneous equation when one solution is already known.
- The method of *undetermined coefficients*, which solves special kinds of inhomogeneous equations with constant coefficients.
- The method of *variation of parameters*, which yields the solution of a general inhomogeneous equation, given a fundamental set of solutions to the homogeneous equation.

Each of these techniques reduces the search for a formula solution to a simpler problem in algebra or calculus: finding the root of a polynomial, computing an antiderivative, or solving a pair of simultaneous linear equations.

The process of finding formulas for exact solutions of equations, either by hand or by computer, is called *symbolic computation*. The **dsolve** command incorporates all of the techniques listed above. It enables us to find exact solutions more rapidly and more reliably than we could by hand.

There are many differential equations that do not yield to the techniques listed above. By using more advanced ideas from calculus, however, we can find exact or approximate solutions for a wider class of differential equations. In this chapter and the next, we discuss two such calculus-based techniques for finding solutions of differential equations: *series solutions* and *Laplace transforms*. The method of series solutions constructs power series solutions of linear differential equations with variable coefficients; in many examples, we compute only finitely many terms of the power series and therefore obtain Taylor polynomial approximations of the solutions. The theory of Laplace transforms (discussed in the next chapter) enables us to solve constant coefficient linear differential equations with discontinuous inhomogeneous terms. These are sophisticated techniques, involving improper integrals, complex variables, and power series.

For each technique, we describe how to use MATLAB to automate the process of computing a symbolic solution to a differential equation or to an associated initial value problem. The methods discussed in these chapters are different from **dsolve**, because they involve a *sequence* of MATLAB commands. As with numerical and graphical methods, computationally intensive symbolic methods become more tractable when done with the aid of a mathematical software system.

11.1 Series Solutions

Consider the second order homogeneous linear differential equation with variable coefficients

$$a(x)y''(x) + b(x)y'(x) + c(x)y(x) = 0. \tag{11.1}$$

Suppose the coefficient functions a, b, and c are *analytic* (*i.e.*, have convergent power series representations) in a neighborhood of a point $x = x_0$. For simplicity, we shall assume that $x_0 = 0$, but everything we say remains valid for any point of analyticity. We begin by dividing by $a(x)$ to normalize the equation. This causes no problems as long as $a(0) \neq 0$. In this case, we refer to the origin as an *ordinary point* for the equation. Changing notation, we write

$$y''(x) + p(x)y'(x) + q(x)y(x) = 0. \tag{11.2}$$

Equations (11.1) and (11.2) have the same solutions. We will search for a power series solution of the form

$$y(x) = \sum_{n=0}^{\infty} a_n x^n. \tag{11.3}$$

If we can find such a solution, then $y(0) = a_0$ and $y'(0) = a_1$. Thus, if we are given initial values, we can find the first two coefficients of the power series solution.

By substituting the infinite series (11.3) into equation (11.2), expanding p and q as power series, and combining terms of the same degree, we obtain a *recursion relation* for the coefficients a_n. In other words, for each $n \geq 2$, we obtain an algebraic equation for a_n involving $a_0, a_1, \ldots, a_{n-1}$ and the coefficients of the power series of p and q. Sometimes, we can solve the recursion relation in *closed form* by finding an algebraic formula for a_n. This formula would involve known functions of n (such as $n!$ or powers of n) and a_0 and a_1, but no other coefficients of lower degree. If no initial data are given, then a solution to the recursion relation produces a general solution of the differential equation with arbitrary constants a_0 and a_1.

Solving a recursion relation in closed form is like finding an exact formula solution for a differential equation. In many cases, we cannot solve the recursion relation. Nevertheless, we can still use the recursion relation to compute as many coefficients as we wish. As long as we stay close to the origin, the leading terms should give a good approximation to the full power series solution. One must be careful, however, because the power series solution will be valid only inside its radius of convergence. The radius of convergence will be at least as large as the distance from the origin to the nearest singularity of p or q. When the recursion relation is solvable in closed form, we can use a standard test from calculus to compute the radius of convergence precisely.

We can use MATLAB to automate the process of finding power series solutions. We shall describe a straightforward technique for computing the leading coefficients explicitly.

Example 11.1 We would like to compute the first five terms in the power series approximation of the initial value problem

$$y'' - xy' - y = 0, \qquad y(0) = 2, \qquad y'(0) = 1. \tag{11.4}$$

We will use function M-files to define new functions, **formalseries** and **sersol**, to find power series solutions. These function M-files are given in Section 11.3. You should enter them in separate function M-files called `formalseries.m` and `sersol.m`, respectively. Make sure that you put the M-files in a place where MATLAB will find them, either in the current working directory or in MATLAB's search path. You need not understand how the M-files work (though we encourage you to try), just how to use them to produce series solutions. These M-files are also available on the internet at `http://www.math.umd.edu/undergraduate/schol/ode/Mfiles`.

Assuming that the M-files have already been set up, as the first step in solving the initial value problem we use the **formalseries** function to create a formal (truncated) power series. We specify the independent variable and the order of the truncated series.

```
>> syms y x
>> y = formalseries(x, 5)
y =
a0 + a1*x + a2*x^2 + a3*x^3 + a4*x^4 + a5*x^5
```

Then we compute the first and second derivatives of the formal power series and substitute them into the differential equation.

```
>> dy = diff(y, x);
>> d2y = diff(dy, x);
>> ode = collect(d2y - x*dy - y, x)
ode =
- 6*a5*x^5 - 5*a4*x^4 + (20*a5 - 4*a3)*x^3
+ (- 3*a2 + 12*a4)*x^2 + ( 2*a1 + 6*a3)*x + 2*a2 - a0
```

Now we use the **sersol** command to calculate the series solution.

```
>> soln = sersol(ode, x, 5, [2 1])
soln =
2 + x + x^2 + 1/3*x^3 + 1/4*x^4 + 1/15*x^5
```

Note that the arguments to **sersol** consist of the **formalseries** expansion of the differential equation (**ode**), the dependent variable (**x**), the desired order (**5**) of the solution (which must be no larger than the order used in **formalseries**), and the vector of initial conditions (**[2 1]**). You can type **help sersol** for a description of the command. If you want to plot or evaluate the solution, you can use **ezplot** or **subs**.

While the **dsolve** command will find exact solutions to many homogeneous second order equations with polynomial coefficients, equations with coefficients that are not polynomials generally require series methods. When **dsolve** does find solutions, they often involve special functions like **besselj** and **bessely**.

11.2 Singular Points

In the discussion above, we considered equations of the form

$$a(x)y'' + b(x)y' + c(x)y = 0,$$

where $a(0) \neq 0$, and we transformed the equation into the form

$$y'' + p(x)y' + q(x)y = 0$$

by dividing by $a(x)$. If $a(0) = 0$, dividing by $a(x)$ may result in one or both of $p(x)$ and $q(x)$ being singular at $x = 0$. In this case, we say that the equation has a *singularity* at $x = 0$.

A prototype is the Euler equation:

$$x^2y'' + bxy' + cy = 0,$$

where b and c are constants. Dividing by x^2 yields the equation

$$y'' + \frac{b}{x}y' + \frac{c}{x^2}y = 0, \tag{11.5}$$

so $p(x) = b/x$ and $q(x) = c/x^2$. Both $p(x)$ and $q(x)$ are singular at $x = 0$. Such isolated singularities, where the singularity of p is no worse than $1/x$ and the singularity of q is no worse than $1/x^2$, are called *regular singular points*. More precisely, we say that a function $f(x)$ has a *pole of order* n at x_0 if $\lim_{x\to x_0} f(x) = \infty$, if $(x - x_0)^n f(x)$ has a convergent power series expansion around $x = x_0$, and n is the smallest integer such that $\lim_{x\to x_0}(x - x_0)^n f(x)$ is finite. A homogeneous linear differential equation $y^{(n)} + p_1(x)y^{(n-1)} + \cdots + p_n(x)y = 0$ is said to have a regular singular point at x_0 if it is singular at x_0, and p_k has a pole of order at most k at x_0.

Note that the Euler equation (11.5) has x^r as a solution, provided that r satisfies the *indicial equation*, $r(r-1) + br + c = 0$. The solution procedure for general second order equations with regular singular points is based on this example as a prototype. We suppose, for simplicity, that the singular point is $x = 0$, and we look for solutions on the interval $(0, \infty)$. Let $p_0 = \lim_{x\to 0} xp(x)$ and $q_0 = \lim_{x\to 0} x^2q(x)$. Let r_1 and r_2 be the roots of the indicial equation, $r(r-1)+p_0r+q_0=0$, with $r_1 \geq r_2$ if the roots are real. We look for a solution of the differential equation of the form

$$y_1(x) = x^{r_1}u(x),$$

where $u(0)=1$. If $r_1 - r_2$ is not an integer, then we look for a second solution of the form

$$y_2(x) = x^{r_2}v(x),$$

where $v(0)=1$. If $r_1 = r_2$, then we look for a second solution of the form

$$y_2(x) = y_1(x)\ln(x) + x^{r_1}v(x),$$

where $v(0)=0$. If $r_1 - r_2$ is a positive integer, then we look for a second solution of the form

$$y_2(x) = ay_1(x)\ln(x) + x^{r_2}v(x),$$

where a is a constant (which might be 0) and $v(0)=1$. The functions $u(x)$ and $v(x)$ will be analytic at 0 and therefore will have power series expansions at $x=0$. These solutions are called *Frobenius series* solutions.

Example 11.2 Consider the differential equation:

$$2xy'' + y' + xy = 0. \tag{11.6}$$

The origin is a regular singular point, with $p_0 = 1/2$ and $q_0 = 0$. The indicial equation is $r(r-1)+r/2=0$, which has roots $r_1 = 1/2$ and $r_2 = 0$. Hence, the differential equation will have one Frobenius series solution that is analytic, and one of the form

$$x^{1/2}\sum_{n=0}^{\infty} a_n x^n,$$

with $a_0 = 1$. The solution is:

$$\begin{aligned} y(x) = {} & C1\,\sqrt{x}\left(1 - \frac{1}{10}x^2 + \frac{1}{360}x^4 + O\left(x^6\right)\right) \\ & + C2\left(1 - \frac{1}{6}x^2 + \frac{1}{168}x^4 + O\left(x^6\right)\right). \end{aligned}$$

Note that we did not specify an initial condition in this example, although we could have since the solutions have finite limits at 0. In general, specifying an initial condition (at the singular point) for a singular differential equation will cause **`dsolve`** to fail.

We will adapt the series solution method for ordinary points to calculate the leading terms of the Frobenius series corresponding to $r_1 = 1/2$. We will define $y(x)$ to be $\sqrt{x}\,u(x)$ and expand both y and the differential equation. Here is the sequence of commands:

```
>> clear all
>> syms u y x
>> u = formalseries(x, 5);
>> y = x^(1/2)*u
y =
x^(1/2)*(a0 + a1*x + a2*x^2 + a3*x^3 + a4*x^4 + a5*x^5)
>> dy = diff(y, x);
>> d2y = diff(dy, x);
```

Here is the series version of the differential equation.

```
>> ode = collect(2*x*d2y + dy + x*y, x)
ode =
a5*x^(13/2) + a4*x^(11/2) + (55*a5 + a3)*x^(9/2)
+ (36*a4 + a2)*x^(7/2) + (21*a3 + a1)*x^(5/2)
+ (10*a2 + a0)*x^(3/2) + 3*a1*x^(1/2)
```

Since the lowest order term in this fractional series is $x^{1/2}$, we must multiply the series by $x^{-1/2}$ to turn it into a bona fide power series. Note that in order for the **sersol** function to work, the resulting power series must have a nontrivial constant term, in this case **3*a1**. If we were to multiply, say, by $x^{1/2}$, then we would obtain a power series with no constant term, and **sersol** would fail.

```
>> ode = collect(simplify(ode*x^(-1/2)), x)
ode =
a5*x^6 + a4*x^5 + (55*a5 + a3)*x^4 + (36*a4 + a2)*x^3
+ (21*a3 + a1)*x^2 + (10*a2 + a0)*x + 3*a1
```

Now we use **sersol** to calculate the series solution. In the case of a Frobenius series for a second order equation, we must specify exactly one initial condition when we use **sersol** (because once **a0** is specified, all the other coefficients are determined by setting **ode = 0**). We use the condition $u(0) = 1$ (which is equivalent to setting **a0** equal to 1).

```
>> soln = sersol(ode, x, 5, [1])
soln =
1 - 1/10*x^2 + 1/360*x^4
```

This is the Taylor series approximation for $u(x)$. To obtain the series for $y(x)$, we multiply by $x^{1/2}$.

```
>> sol1 = collect(simplify(x^(1/2)*soln), x)
sol1 =
1/360*x^(9/2) - 1/10*x^(5/2) + x^(1/2)
```

To obtain another linearly independent solution of the equation, we repeat this procedure using the other exponent, $r_2 = 0$. This second solution will be analytic, so we can use the procedure described in the first section of this chapter, with one modification: we must specify only one initial condition in the last argument to **sersol** (instead of using two, as we did for nonsingular equations).

```
>> clear all
>> syms x y
>> y = formalseries(x, 5);
```

```
>> dy = diff(y, x);
>> d2y = diff(dy, x);
```

Here's the series version of the ode.

```
>> ode = collect(2*x*d2y + dy + x*y, x)
ode =
a5*x^6 + a4*x^5 + (45*a5 + a3)*x^4 + (28*a4 + a2)*x^3
+ (15*a3 + a1)*x^2 + (6*a2 + a0)*x + a1
```

Now we use **sersol** to solve for the coefficients, with the condition $y(0) = 1$.

```
>> soln = sersol(ode, x, 5, [1])
soln =
1 - 1/6*x^2 + 1/168*x^4
```

This is the Taylor series approximation for a second linearly independent solution $y(x)$.

Exercise 11.1 Compute two more terms in each of the examples presented in this chapter.

It is important to note that for equations with regular singular points, the Frobenius series tells us how fast the solution blows up at the singularity; *e.g.*, the solution blows up like $1/x$, $x^{-1/2}$, *etc*. This information is not easily gleaned from a numerical solution. For equations with an irregular singular point, we cannot expect a Frobenius series solution to be valid, and other techniques (numerical or qualitative) must be used. Some of those techniques are addressed in Problem Set E.

In this chapter, we have focused on linear homogeneous second order differential equations. The method of series solutions can also be used for inhomogeneous equations, higher order equations, and nonlinear equations. As we've noted before, most differential equations cannot be solved in terms of elementary functions. For many of these equations, a series solution is the only formula solution available.

11.3 Function M-files for Series Solutions

Here are the contents of two function M-files for computing series solutions. You should enter these in separate files called `formalseries.m` and `sersol.m`. You can also download these files from

`http://www.math.umd.edu/undergraduate/schol/ode/Mfiles/`.

formalseries.m

```
function u = formalseries(x, n, x0)

% Creates a formal power series.
%
% FORMALSERIES(x) returns the terms up to degree 6 of
% a formal power series in the variable x, centered
% at x = 0.
```

```
%
% FORMALSERIES(x, n) returns the terms up to degree n
% of the formal power series in x at 0.
%
% FORMALSERIES(x, n, x0) returns the terms up to degree
% n of the formal power series in x centered at x = x0.

if nargin < 3
  x0 = 0;
end
if nargin < 2
  n = 6;
end
x = sym(x);

% The next loop builds a formal polynomial of degree n
% in the variable x centered at the point x0.
u = sym(0);
for k = 0:n
  v = sym(['a' num2str(k)]);
  u = u + v*(x - x0)^k;
end
```

sersol.m

```
function soln = sersol(ode, x, order, initial, x0)

% Series solution of ordinary differential equation.
%
% SERSOL(ode, x, order, initial, x0) returns the series
% solution up to the specified order (at x = x0) of the
% ordinary differential equation ode with independent
% symbolic variable x.  The fourth argument is a vector
% of initial conditions, starting with y(0), y'(0), ....
% For nonsingular equations, the number of initial
% conditions must equal the order of the equation.  If
% the equation has a regular singular point at 0, then
% only one initial condition, representing the first
% coefficient in the series, should be given.  The fifth
% argument can be omitted, in which case x0 = 0 is used.
%
% The first argument, ode, is a symbolic expression or
% string containing a formal Taylor polynomial expansion
```

```
% of the differential equation with nonzero constant
% term.  For a nonsingular equation, ode should be the
% result of substituting the expression y = a0 +
% a1*(x - x0) + a2*(x - x0)^2 + ..., which can be created
% with the function FORMALSERIES, into the equation.  If
% the equation has a regular singular point at 0, then
% the Frobenius series y = (x - x0)^r*(a0 + a1*(x - x0)
% + a2*(x - x0)^2 + ...)  (for appropriate r) should be
% substituted instead, and before using SERSOL, the
% result must be multiplied or divided by an appropriate
% power of (x - x0) and simplified to make it a
% polynomial that is nonzero at x0.  In this case, the
% output of SERSOL represents the analytic part of the
% series for y; it should be multiplied by (x - x0)^r to
% get the series solution.
%
% Example:
% syms y x;
% y = formalseries(x, 5)
% dy = diff(y, x);
% d2y = diff(dy, x);
% ode = collect(d2y - x*dy - y, x)
% soln = sersol(ode, x, 5, [1, 0])

% Error checking.
if nargin < 4
  error('Not enough input  arguments.')
end

% Initialization.
if nargin < 5
  x0 = 0;
end
ode = sym(ode);
x = sym(x);
x0 = sym(x0);
len = length(initial);

% Set coefficients determined by the initial conditions.
for k = 0:(len-1)
  vars{k+1} = ['a' num2str(k)];
  vals{k+1} = sym(initial(k+1));
end
```

```
% Create the system of equations.
for k = len:degree
  eqn = subs(diff(ode, x, k - len), x, x0);
  eqn = subs(eqn, {vars{1:len}}, {vals{1:len}});
  eqns{k+1} = [char(eqn) '=0'];
  vars{k+1} = ['a' num2str(k)];
end

% Solve the equations for the unknown coefficients.
range = (len+1):(degree+1);
sol = solve(eqns{range}, vars{range});
for k = range
  vals{k} = sol.(vars{k});
end

% Substitute the coefficients into a formal series
soln = subs(formalseries(x, degree, x0), vars, vals);
```

11.4 Series Solutions Using `maple`

If you are using the Professional Version of MATLAB (*not* the Student Version), then it is possible to obtain series solutions of differential equations (without using the M-files above) using the powerful symbolic capabilities of the Maple kernel. The idea is to use features of Maple's **dsolve** command that are not available in the MATLAB version of the same command. The command **mhelp dsolve** will display details of these features and the exact syntax of Maple's **dsolve** command (unless you are using the Student Version, in which case it will give an error message). Without going into full details, we give a few examples.

Example 11.3 Consider the homogeneous linear differential equation

$$t^2y'' + y = 0. \tag{11.7}$$

One can find the complete series expansion of the most general solution using the Maple kernel as follows:

```
>> maple('dsolve', 't^2*diff(diff(y(t),t),t) + y(t)', ...
     'y(t)', 'formal_series')
ans =
y(t) = 1/gamma(3/2+1/2*i*3^(1/2))/pi*Sum(
(-_C1*gamma(n-1/2+1/2*i*3^(1/2))*pi
+_C2*sin(1/2*pi*(-1+i*3^(1/2)))*gamma(3/2+1/2*i*3^(1/2))^2
*gamma(n-1/2-1/2*i*3^(1/2)))*(t+1)^n/n!,n = 0 .. Inf)
```

This is Maple's way of saying that the solution can be written in the form

$$y(t) = \frac{1}{\pi\Gamma\left(\frac{3+i\sqrt{3}}{2}\right)} \sum_{n=0}^{\infty} \Bigg(-C_1\Gamma\left(n - \frac{1-i\sqrt{3}}{2}\right)\pi$$
$$+ C_2 \sin\left(\frac{\pi}{2}\left(-1+i\sqrt{3}\right)\right)\Gamma\left(\frac{3+i\sqrt{3}}{2}\right)^2 \Gamma\left(n - \frac{1+i\sqrt{3}}{2}\right)\Bigg) \frac{(t+1)^n}{n!}.$$

Of course, it is not clear that this is terribly useful, so one might prefer simply to explicitly compute the first few terms of the solution. One can do this with the option **series** in place of the option **formal_series**:

```
>> maple('dsolve', 't^2*diff(diff(y(t),t),t) + y(t)', ...
    'y(t)', 'series')
y(t) = _C1*t^(1/2-1/2*i*3^(1/2))*(series(1+O(t^6),t,6))
+_C2*t^(1/2+1/2*i*3^(1/2))*(series(1+O(t^6),t,6))
```

This indicates that the solution has the form

$$y(t) = C_1 t^{\frac{1-i\sqrt{3}}{2}} + C_2 t^{\frac{1+i\sqrt{3}}{2}}, \tag{11.8}$$

plus terms of higher order in t. Equation (11.7) can be solved, say with MATLAB's **dsolve**; the solution is

$$y(t) = C_1 \sqrt{t} \sin\left(1/2\sqrt{3}\ln(t)\right) + C_2 \sqrt{t} \cos\left(1/2\sqrt{3}\ln(t)\right),$$

which, when one disentangles the meaning of the complex exponentials, is seen to be equivalent to (11.8). Thus the series starting with (11.8) does not, in fact, have any higher order terms.

Example 11.4 Here's another example with an initial value problem. Suppose we want to solve the nonlinear differential equation

$$y' + ty^2 = \sin(t), \quad y(0) = 0. \tag{11.9}$$

One can't solve this exactly by any standard method, so we look for a series solution.

```
>> maple('dsolve', ...
    '{diff(y(t), t) + t*y(t)^2 = sin(t), y(0) = 0}', ...
    'y(t)', 'series')
ans =
y(t) = series(1/2*t^2-1/24*t^4+O(t^6),t,6)
```

That means the solution is $\frac{t^2}{2} - \frac{t^4}{24}$ plus terms starting with a constant times t^6, and so is approximately $\frac{t^2}{2} - \frac{t^4}{24}$ for small t. To set more terms in the series, we need to reset the Maple kernel's internal variable **Order** as follows:

```
>> maple('Order := 8');
>> maple('dsolve', ...
    '{diff(y(t), t) + t*y(t)^2 = sin(t), y(0) = 0}', ...
    'y(t)', 'series')
ans =
y(t) = series(1/2*t^2-1/24*t^4-29/720*t^6+O(t^8),t,8)
```

Thus the solution has the form $\frac{t^2}{2} - \frac{t^4}{24} - \frac{29t^6}{720}$ plus terms starting with a constant times t^8.

Chapter 12

Laplace Transforms

A *transform* is a mathematical operation that changes a given function into a new function. Transforms are often used in mathematics to change a difficult problem into a more tractable one. In this chapter, we introduce the *Laplace Transform*, which is particularly useful for solving linear differential equations with constant coefficients and discontinuous inhomogeneous terms. The key feature of the Laplace Transform is that (roughly speaking) it changes the operation of differentiation into the operation of multiplication. Thus the Laplace Transform changes a differential equation into an algebraic equation. To solve a linear differential equation with constant coefficients, you apply the Laplace Transform to change the differential equation into an algebraic equation, solve the algebraic equation, and then apply the Inverse Laplace Transform to transform the solution of the algebraic equation back into the solution of the differential equation.

The Laplace Transform of a function f is a new function, denoted by F or by $\mathcal{L}(f)$, and defined as follows:

$$F(s) = \mathcal{L}(f)(s) = \int_0^\infty f(t)e^{-st}\,dt.$$

This transform is called an *integral transform* because it is obtained by integrating the function f against another function e^{-st}, called the *kernel* of the transform. The integral in question is an improper integral, so we have to make sure that it converges. Notice that while the argument s of the function F appears as a parameter in the integrand, the integration is with respect to the variable t. Also, the integral is over the domain $[0, \infty)$, so we assume that the function f is defined for $t \geq 0$.

To get a feel for the Laplace Transform, we compute the Laplace Transform of the function $g(t) = e^{at}$.

$$\begin{aligned}
\mathcal{L}(g)(s) &= \int_0^\infty e^{at}e^{-st}\,dt = \lim_{b\to\infty}\int_0^b e^{at}e^{-st}\,dt \\
&= \lim_{b\to\infty}\int_0^b e^{(a-s)t}\,dt = \lim_{b\to\infty}\left.\frac{e^{(a-s)t}}{(a-s)}\right|_0^b
\end{aligned}$$

$$
\begin{aligned}
&= \lim_{b\to\infty}\left(\frac{e^{(a-s)b}}{(a-s)} - \frac{1}{a-s}\right) \\
&= \begin{cases} 1/(s-a), & \text{if } s > a \\ +\infty, & \text{if } s \le a. \end{cases}
\end{aligned}
$$

To avoid the infinite values, we say that the Laplace Transform of e^{at} is defined only for $s > a$. A straightforward argument shows that if f is any piecewise continuous function on $[0, \infty)$ with the property that $|f(t)| \le Ke^{at}$, for some constant $K > 0$, then the improper integral defining the Laplace Transform converges for $s > a$, and therefore $\mathcal{L}(f)(s)$ is defined at least for $s > a$. (We say that a function is *piecewise continuous* if it only has a discrete set of jump discontinuities.) If f satisfies an inequality of the form $|f(t)| \le Ke^{at}$, we say that f is of *exponential order*. Most functions one encounters in practice are of exponential order. In the rest of this chapter, we only consider piecewise continuous functions of exponential order. In particular, if f is a bounded function, then it satisfies the inequality $|f(t)| \le K = Ke^{0t}$ for some $K > 0$, so $\mathcal{L}(f)(s)$ is defined for all $s > 0$. More generally, any function whose growth is of exponential order has a Laplace Transform which is defined for sufficiently large s.

Exercise 12.1 Compute the Laplace Transforms of the functions $f(t) = 1$ and $\cos t$. (For $\cos t$, you must integrate by parts twice.)

We asserted that the Laplace Transform changes differentiation into multiplication. This is a consequence of the integration by parts formula:

$$
\begin{aligned}
\mathcal{L}(f')(s) &= \int_0^\infty f'(t)e^{-st}\,dt = \lim_{b\to\infty}\int_0^b f'(t)e^{-st}\,dt \\
&= \lim_{b\to\infty}\left[f(t)e^{-st}\big|_0^b - \int_0^b f(t)(-se^{-st})\,dt\right] \\
&= -f(0) + s\lim_{b\to\infty}\int_0^b f(t)e^{-st}\,dt \\
&= s\mathcal{L}(f)(s) - f(0). \qquad (12.1)
\end{aligned}
$$

Here we have assumed that f is differentiable and f' is piecewise continuous and of exponential order. We can summarize (12.1) as follows: If the Laplace Transform of f is $F(s)$, then the Laplace Transform of f' is $sF(s) - f(0)$. In other words, the Laplace Transform changes the operation of differentiation into the operation of multiplication (by the independent variable) plus a translation (by $-f(0)$).

Remark 12.1 If you know $\mathcal{L}(f')(s)$ and $f(0)$, you can add them and divide by s to obtain $\mathcal{L}(f)(s)$. This procedure is analogous to solving a first order initial value problem by integration. You may have noticed in Chapter 9 that the icon for an Integrator block in Simulink contains the expression $1/s$. Because many engineers and scientists use the Laplace Transform to analyze differential equations, Simulink uses "division by s" as the mnemonic for integration.

Applying the formula in (12.1) repeatedly yields the following generalization to higher derivatives. If the Laplace Transform of f is F, $f^{(k)}$ is continuous for $k = 0 \ldots n-1$, and $f^{(n)}$ is piecewise continuous, then

$$\mathcal{L}(f^{(n)})(s) = s^n F(s) - s^{n-1} f(0) - s^{n-2} f'(0) - \cdots - f^{(n-1)}(0).$$

The Laplace Transform has many other important properties, of which we mention three here. First, the Laplace Transform is invertible, in the sense that knowing the function $\mathcal{L}(f)(s)$ determines $f(t)$ for $t \geq 0$, except at points where it is discontinuous. (This is as good as we can expect, since changing f for $t < 0$ does not affect $\mathcal{L}(f)$, nor does changing its value at a point of discontinuity.) The Inverse Laplace Transform is denoted $\mathcal{L}^{-1}$. Like the Laplace Transform, the Inverse Laplace Transform can be written as an integral transform, but it involves contour integrals in the complex plane, so we do not give the definition here. Second, the Laplace Transform is linear; *i.e.*, $\mathcal{L}(af + bg) = a\mathcal{L}(f) + b\mathcal{L}(g)$, where a and b are constants. Linearity of the Laplace Transform follows easily from linearity of integration. Third, the Inverse Laplace Transform is linear. Linearity of the Inverse Laplace Transform follows from the linearity of the Laplace Transform.

12.1 Differential Equations and Laplace Transforms

Let's see what happens when we apply the Laplace Transform to a second order linear differential equation with constant coefficients. Consider the initial value problem

$$ay''(t) + by'(t) + cy(t) = f(t), \qquad y(0) = y_0, \qquad y'(0) = y_0'.$$

If we apply the Laplace Transform to this equation and use the initial conditions, we get the algebraic equation

$$a(s^2 Y(s) - s y_0 - y_0') + b(sY(s) - y_0) + cY(s) = F(s),$$

where $Y(s)$ is the Laplace Transform of $y(t)$ and $F(s)$ is the Laplace Transform of $f(t)$. We solve this algebraic equation for $Y(s)$ to get

$$Y(s) = \frac{F(s) + a s y_0 + a y_0' + b y_0}{a s^2 + b s + c},$$

and then compute the Inverse Laplace Transform of the right-hand side to get an expression for $y(t)$. The resulting solution is only guaranteed to be valid for $t \geq 0$, since the Laplace Transform method only takes into account how $f(t)$ behaves for $t \geq 0$.

Remark 12.2 In control theory, a common problem is to find a forcing function $f(t)$ that will make the solution $y(t)$ behave in a desired fashion. Having taken the Laplace Transform of the differential equation, this problem becomes algebraic: find $F(s)$ that makes $Y(s)$ have a desired form. One must work with Laplace Transforms for a while to develop an understanding of how the form of $Y(s)$ determines the behavior of the solution $y(t)$.

But having developed this sense, many engineers and scientists find it helpful to think in terms of the Laplace Transform of a differential equation that models a system rather than think directly about the differential equation. While the equations we consider in this chapter can be solved directly using **dsolve**, one purpose of the examples in this section and the corresponding problems in Problem Set E is for you to see how the solutions and their Laplace Transforms correspond.

Traditionally, one used tables to look up Laplace Transforms and Inverse Laplace Transforms. But you can use the MATLAB commands **laplace** and **ilaplace** to compute Laplace Transforms and Inverse Laplace Transforms. For example, to compute the Laplace Transform of $\cos t$, type:

```
>> syms s t; laplace(cos(t), t, s)
ans =
s/(s^2+1)
```

The second and third input arguments to **laplace** specify the independent variables of the original function and the transformed function; they can be omitted if you use the usual variables t and s. To compute the Inverse Laplace Transform of the output above, type:

```
>> ilaplace(s/(s^2 + 1), s, t)
ans =
cos(t)
```

Exercise 12.2 Compute the Laplace Transform of t^2 and e^{at}. (Before computing the Laplace Transform of e^{at}, declare **a** to be a symbolic variable with **syms a**.) Compute the Inverse Laplace Transform of $1/(s^2 - 1)$.

We can use these commands to implement the Laplace Transform method for solving differential equations. Consider the initial value problem $y'' + y = \sin 2t$, $y(0) = 1$, $y'(0) = 0$. First, we define the equation that we want to solve and compute its Laplace Transform:

```
>> eqn = sym('D(D(y))(t) + y(t) = sin(2*t)');
>> lteqn = laplace(eqn, t, s)
lteqn =
s*(s*laplace(y(t),t,s)-y(0))-D(y)(0)+laplace(y(t),t,s) = 2/(s^2+4)
```

Remark 12.3 The commands **laplace** and **ilaplace** are part of the Symbolic Math Toolbox, which as we have mentioned before is based on another software package called Maple. In the input above, we used the Maple notation **D(D(y))(t)** for $y''(t)$, and in the output we see **D(y)(0)**, which means $y'(0)$.

To simplify matters, we will replace the expression **laplace(y(t),t,s)** in **lteqn** with **Y**. At the same time, let us substitute the initial values 1 and 0 for the expressions **y(0)** and **D(y)(0)**, respectively:

```
>> syms Y; neweqn = subs(lteqn, {'laplace(y(t),t,s)', ...
      'y(0)', 'D(y)(0)'}, {Y, 1, 0})
```

```
neweqn =
s*(s*Y-1)+Y=2/(s^2+4)
```

(You should take care not to put in any spaces when typing **'laplace(y(t),t,s)'** above, because it appears without spaces in **lteqn**.) Now we solve this equation for **Y**:

```
>> ytrans = solve(neweqn, Y)
ytrans =
(s^3+4*s+2)/(s^2+4)/(s^2+1)
```

Finally, we apply the Inverse Laplace Transform to **ytrans** to obtain the solution of the initial value problem:

```
>> y = ilaplace(ytrans, s, t)
y =
cos(t)+2/3*sin(t)-1/3*sin(2*t)
```

You can evaluate or plot this solution by using **subs** or **ezplot**, like in Example 5.1. For example, to plot the solution over the interval $[0, 5]$, type:

```
>> ezplot(y, [0 5])
```

The resulting graph appears in Figure 12.1.

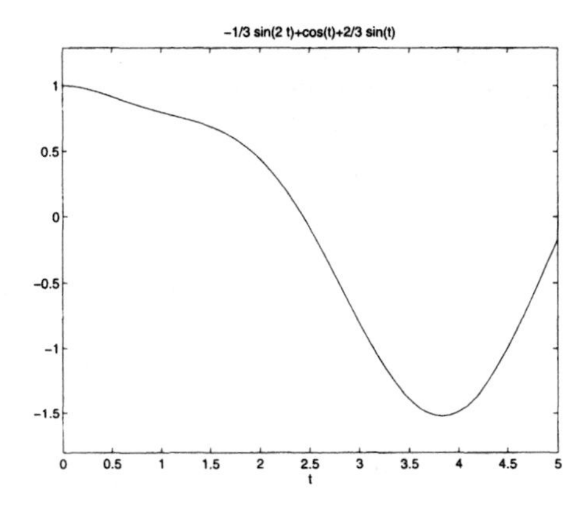

Figure 12.1: Laplace Transform Solution of $y'' + y = \sin 2t$, $y(0) = 1$, $y'(0) = 0$

You can follow the steps above to solve other linear equations with constant coefficients. The only things you need to change are the definition of the equation and the values you substitute for the initial conditions.

Example 12.1 Consider the initial value problem:

$$y''(t) + 2y'(t) = e^{-t}, \quad y(0) = 1, \quad y'(0) = 2.$$

Here are the MATLAB commands for producing a solution.

```
>> syms s t Y
>> eqn = sym('D(D(y))(t) + 2*D(y)(t) = exp(-t)');
```

```
>> lteqn = laplace(eqn, t, s);
>> neweqn = subs(lteqn, {'laplace(y(t),t,s)', ...
       'y(0)', 'D(y)(0)'}, {Y, 1, 2});
>> ytrans = solve(neweqn, Y);
>> y = ilaplace(ytrans, s, t)
y =
5/2-exp(-t)-1/2*exp(-2*t)
```

12.2 Discontinuous Functions

The Laplace Transform is especially useful for solving differential equations that involve piecewise continuous functions. The basic building block for piecewise continuous functions is the *unit step function* $u(t)$, defined by

$$u(t) = \begin{cases} 0, & \text{if } t < 0, \\ 1, & \text{if } t \geq 0. \end{cases}$$

In honor of Oliver Heaviside (1850–1925), who developed the Laplace Transform method to solve problems in electrical engineering, this function is sometimes called the Heaviside function. In MATLAB's Symbolic Math Toolbox, it is called **heaviside**.

The unit step function is best thought of as a switch, which is off until time 0 and then is on starting at time 0. To make a switch that comes on at time c, we simply translate by c; thus $u(t - c)$ is a switch that comes on at time c. Similarly, $1 - u(t - c)$ is a switch that goes off at time c. The function $u(t - c)$ is sometimes written $u_c(t)$.

The unit step function can be used to build piecewise continuous functions by switching pieces of the function on and off at appropriate times. Consider, for example, the function

$$f(t) = \begin{cases} 0, & \text{if } t < 0, \\ 1, & \text{if } 0 \leq t < 1, \\ t^2, & \text{if } 1 \leq t < 3, \\ \sin 2t, & \text{if } t \geq 3. \end{cases}$$

We can write this as

$$f(t) = 0 + u(t)(1 - 0) + u(t - 1)(t^2 - 1) + u(t - 3)(\sin 2t - t^2).$$

We started with the formula for $f(t)$, in this case 0, that is valid for the leftmost values of t. Then for each point c where the formula for $f(t)$ changes, in this case $c = 0, 1, 3$, we added the term

$$u(t - c)(\text{formula to the right of } c - \text{formula to the left of } c).$$

In other words, at each value of c we switched on the difference between the formula for $f(t)$ that is valid just to the right of c and the formula that is valid just to the left of c. In MATLAB, having declared **t** to be symbolic, you can enter $f(t)$ as

```
>> f = heaviside(t) + heaviside(t - 1)*(t^2 - 1) ...
       + heaviside(t - 3)*(sin(2*t) - t^2);
```

To plot **f**, type:

```
>> ezplot(f, [0,10])
>> title '', axis auto
```

The result appears in Figure 12.2. In this graph, you can clearly see the three different "pieces" of the function. (The vertical segment in the graph is an artifact of MATLAB's plotting routine and is not part of the function.)

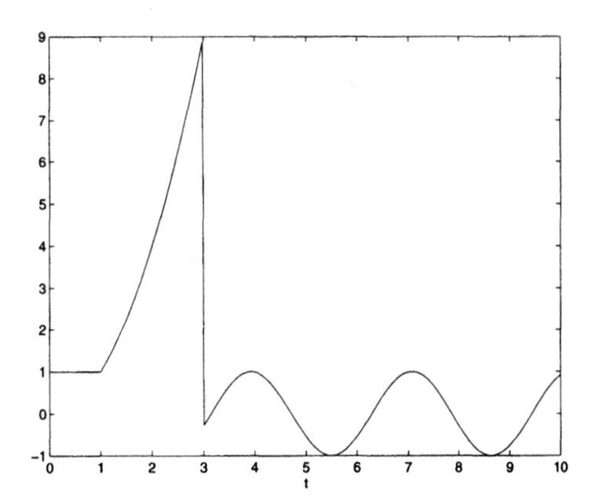

Figure 12.2: A Piecewise Continuous Function

You can also evaluate **f** at any point except for $t = 0, 1, 3$:

```
>> subs(f, t, [1 2])
ans =
   NaN        4
```

Even though we defined $f(t)$ and $u(t)$ for all real t, in MATLAB **heaviside(0)** is undefined, resulting above in the expression **NaN**, which stands for "Not a Number".

Another important discontinuous function is the *Dirac delta function*, usually denoted $\delta(t)$, and sometimes called the *unit impulse function*. The delta function belongs to a class of mathematical objects called *generalized functions* (it is not a true function). By definition, $\delta(t) = 0$ for $t \neq 0$, but $\int_{-\infty}^{\infty} \delta(t)dt = 1$. There is no finite value of $\delta(0)$ that can make its integral nonzero, so we must regard $\delta(0)$ as infinite.

The motivation for the Dirac delta function is as follows. In physics, the total impulse imparted by a varying force $F(t)$ is the integral of $F(t)$. If we consider a force of total impulse 1 and make this force act over a smaller and smaller time interval around 0, then in the limit we obtain $\delta(t)$. Thus the delta function represents an idealized force of total impulse 1 concentrated at the instant $t = 0$. The magnitude and timing of the impulse can be changed by multiplying the delta function by a constant and translating it. Thus the function $10\delta(t-8)$ represents a force of total impulse 10 concentrated at the instant $t = 8$.

The function $\delta(t-c)$ is sometimes denoted $\delta_c(t)$. In MATLAB's Symbolic Math Toolbox, the delta function is called **dirac**.

Exercise 12.3 Plot the Dirac delta function on the interval $[-1,1]$ (you won't see the impulse at 0). Evaluate the function at 0 and 1. Integrate the function over the interval $[-1,1]$.

We consider now the Laplace Transforms of $u_c(t)$ and $\delta_c(t)$. Since the Laplace Transform of a function depends only on its values for $t \geq 0$, we consider only the case when $c \geq 0$ (otherwise these functions are just 1 and 0, respectively, for $t \geq 0$). Then

$$\mathcal{L}(u_c)(s) = \int_0^\infty u_c(t)e^{-st}\,dt = \int_c^\infty e^{-st}\,dt = \frac{e^{-cs}}{s}, \quad \text{for } s > 0.$$

Strictly speaking, taking the Laplace Transform of $\delta_c(t)$ requires the theory of generalized functions, which we alluded to above. Here is a rough explanation. First, we think of $\delta_c(t)$ as a limit of *bona fide* functions. The easiest way to do this is to use step functions. We can write a function of total integral 1, concentrated on the interval $[c-\varepsilon, c+\varepsilon]$, as

$$\frac{1}{2\varepsilon}(u_{c-\varepsilon}(t) - u_{c+\varepsilon}(t)) = \frac{1}{2\varepsilon}(u(t-c+\varepsilon) - u(t-c-\varepsilon)).$$

Thus in a certain sense,

$$\delta_c(t) = \lim_{\varepsilon \to 0+} \frac{1}{2\varepsilon}(u(t-c+\varepsilon) - u(t-c-\varepsilon)) = u'(t-c) = u_c'(t).$$

The rule (12.1) for the Laplace Transform of a derivative can be justified in the case that the derivative is a generalized function, so for $c > 0$,

$$\mathcal{L}(\delta_c)(s) = s\mathcal{L}(u_c)(s) - u_c(0) = e^{-cs}.$$

Exercise 12.4 Verify that **laplace** correctly computes the Laplace Transforms of the functions **heaviside(t - 2)** and **dirac(t - 3)**.

12.3 Differential Equations with Discontinuous Forcing

Consider an inhomogeneous, second order linear equation with constant coefficients:

$$ay''(t) + by'(t) + cy(t) = g(t).$$

The function $g(t)$ is called the forcing function of the differential equation, because in many physical models $g(t)$ corresponds to the influence of an external force. If $g(t)$ is piecewise continuous or involves the delta function, then we can solve the equation by the method of Laplace Transforms.

Example 12.2 Consider the initial value problem:

$$y''(t) + 3y'(t) + y(t) = g(t), \quad y(0) = 1, \quad y'(0) = 1,$$

where

$$g(t) = \begin{cases} 0, & \text{if } t < 0, \\ 1, & \text{if } 0 \leq t < 1, \\ -1, & \text{if } 1 \leq t < 2, \\ 0, & \text{if } t \geq 2. \end{cases}$$

We can solve this equation in MATLAB with the following sequence of commands.

```
>> syms s t Y
>> g = ['heaviside(t) + heaviside(t - 1)*(-2)' ...
       '+ heaviside(t - 2)'];
>> eqn = sym(['D(D(y))(t) + 3*D(y)(t) + y(t) = ' g]);
>> lteqn = laplace(eqn, t, s);
>> neweqn = subs(lteqn, {'laplace(y(t),t,s)', ...
       'y(0)', 'D(y)(0)'}, {Y, 1, 1});
>> ytrans = solve(neweqn, Y);
>> y = ilaplace(ytrans, s, t)
```

(Since the formula for the equation was too long to type on one line, we used string concatenation, as described in Section 8.1.1, to define it over three lines. The formula for the solution is even longer, so we do not display it here.) Figure 12.3 is a graph of the solution, produced with **ezplot**.

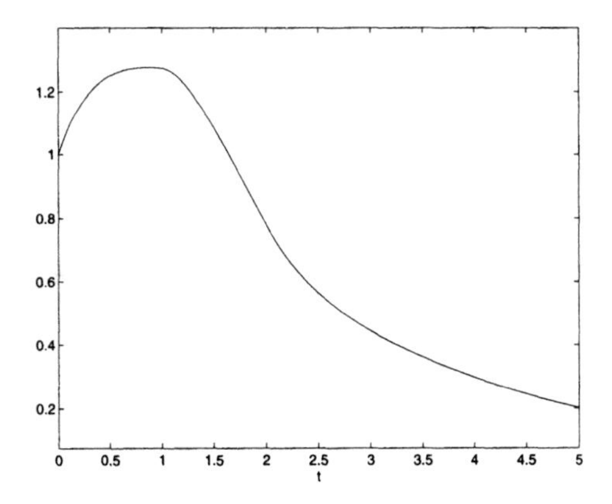

Figure 12.3: Solution of an Equation with Piecewise Continuous Data

Even though the forcing function $g(t)$ is discontinuous at $t = 2$, the solution $y(t)$ is a continuous function. The same is true of solutions of any second order linear differential equation with a discontinuous forcing function. In fact, as long as the forcing function is bounded, the solutions are differentiable too. If the forcing function involves the Dirac delta function, the solutions will be continuous but not differentiable at the times of the impulses, as we will see in the next example.

Example 12.3 We can also use MATLAB to solve inhomogeneous equations involving the delta function. Consider the initial value problem

$$y''(t) + y'(t) + y(t) = \delta_1(t), \quad y(0) = 0, \quad y'(0) = 0.$$

We can solve this equation with the following sequence of commands.

```
>> syms s t Y
>> eqn = sym('D(D(y))(t) + D(y)(t) + y(t) = dirac(t - 1)');
>> lteqn = laplace(eqn, t, s);
>> neweqn = subs(lteqn, {'laplace(y(t),t,s)', ...
            'y(0)', 'D(y)(0)'}, {Y, 0, 0});
>> ytrans = solve(neweqn, Y);
>> y = ilaplace(ytrans, s, t)
y =
2/3*heaviside(t-1)*3^(1/2)*exp(-1/2*t+1/2)*sin(1/2*3^(1/2)*(t-1))
```

Figure 12.4 contains a graph of the solution. In this graph, you can clearly see the effect of the unit impulse at time $t = 1$. Until $t = 1$, the solution is 0. At time $t = 1$, the unit impulse instantaneously changes the velocity y' to 1, though the solution remains continuous. After $t = 1$, the impulse function has no further effect; it is as if we solved the homogeneous equation $y'' + y' + y = 0$ with initial conditions $y(1) = 0$ and $y'(1) = 1$ for $t > 1$. Since the roots of the characteristic polynomial are complex with negative real parts, the solution decays to 0 in an oscillatory manner.

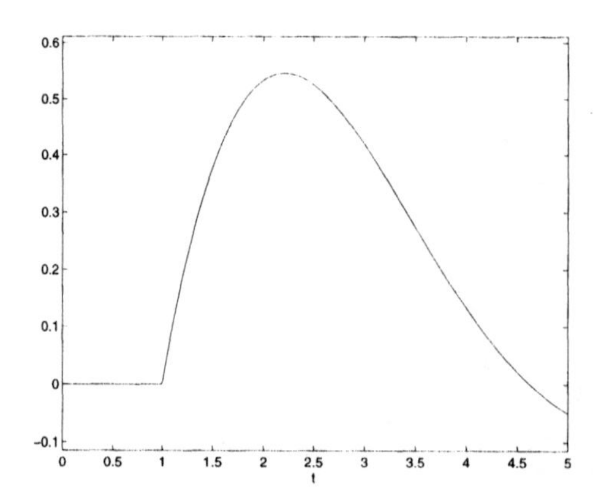

Figure 12.4: Solution of an Equation with an Impulse

Problem Set E

Series Solutions and Laplace Transforms

Problems 1–9 concern series solutions, while Problems 10–21 use Laplace Transforms and/or consider systems with discontinuous forcing functions. The solution to Problem 1 appears in the *Sample Solutions*.

1. Consider Airy's equation

$$y'' - xy = 0.$$

   (a) Compute the terms of degree 10 or less in the Taylor series expansion of the solution $y(x)$ to Airy's equation $y'' - xy = 0$ that satisfies $y(0) = 1$, $y'(0) = 0$.

   (b) On the intervals $(0, 5)$ and $(-10, 0)$, graph the Taylor polynomial you obtained, together with the exact solution of the initial value problem. (You can compute the exact solution with **dsolve**. The solution will be given in terms of the special functions **AiryAi**, **AiryBi**, and **gamma**.) Where is the approximation accurate? (*Hint*: You may have to restrict the range on the plot over negative values.)

   (c) Airy's equation can be written as $y'' = xy$. If $x > 1$, then $y'' = xy > y$, so the solutions of Airy's equation grow at least as quickly as the solutions of $y'' = y$. Analyze the latter equation, and predict how well Taylor polynomials will approximate exact solutions of Airy's equation. Relate your analysis to the graph in part (b).

   (d) Recall from Problem 1 in Problem Set D that the behavior of the solutions to Airy's equation for negative values is oscillatory. Relate that to your graph in part (b) and explain why it fails to be accurate beyond a certain point.

2. Compute the terms of degree 10 or less in the Taylor series expansion of the solution to the following differential equation (*cf.* Problem 23, Section 5.2 of Boyce & DiPrima):

$$y'' - xy' - y = 0, \quad y(0) = 1, \quad y'(0) = 0. \tag{E.1}$$

(a) Calculate and graph the series solution—Taylor polynomial—of degree 10.

(b) What is the long-term behavior of the Taylor polynomial (as $x \to \pm\infty$)? Is the series solution even, odd, or neither?

(c) Find the general solution of the differential equation in (E.1) using **dsolve**, and then find values of the constants so that the initial conditions in (E.1) are satisfied. What is the formula for the solution of (E.1)? Are your conclusion in part (b) confirmed by this formula?

(d) Use **taylor** to calculate the Taylor expansion of the formula obtained in part (b). Does this Taylor expansion agree with the series solution in part (a)?

3. Consider the initial value problem (*cf.* Problem 18, Section 5.2 of Boyce & DiPrima):

$$(1-x)y'' + xy' - y = 0, \quad y(0) = -3, \quad y'(0) = 2.$$

(a) Compute the terms of degree 10 or less in the Taylor series expansion of the solution.

(b) The theory suggests that the true solution may have a singularity at $x = 1$. Graph the Taylor polynomial you obtained. Do you see any signs of the singularity?

(c) Compute the exact solution using **dsolve**. Does the solution have a singularity?

(d) Compute the general solution of the differential equation and discuss its behavior at $x = 1$.

(e) Find the first few terms in the Taylor series expansion at $x = 1$ for two linearly independent solutions of the equation.

4. Use **dsolve** to solve the following Euler equations (*cf.* Section 5.5 of Boyce & DiPrima, Problems 3, 6, 8, 14):

(a) $x^2y'' - 3xy' + 4y = 0,$

(b) $(x-1)^2y'' + 8(x-1)y' + 12y = 0,$

(c) $2x^2y'' - 4xy' + 6y = 0,$

(d) $4x^2y'' + 8xy' + 17y = 0, \quad y(1) = 2, \quad y'(1) = -3.$ In this part, graph the solution and describe how it behaves as $x \to 0$.

5. For the following differential equations, compute the indicial equation and find its roots r_1 and r_2. Then compute the terms of degree 10 or less in the Frobenius series solution corresponding to the larger root. If $r_1 - r_2$ is not an integer, do the same for the other root (*cf.* Problems 1, 5, 8 of Section 5.6 of Boyce & DiPrima):

(a) $2xy'' + y' + xy = 0,$

(b) $3x^2y'' + 2xy' + x^2y = 0,$

(c) $2x^2y'' + 3xy' + (2x^2 - 1)y = 0.$

6. This problem concerns Bessel's equation, which arises in many physical problems with circular symmetry, such as the study of water waves in a circular pond, vibrations of a circular drum, or the diffraction of light by the circular aperture of a telescope. The Bessel function $J_n(x)$ (where the n is an arbitrary integer) is defined to be the coefficient of t^n in the power series expansion of the function

$$\exp\left(\frac{1}{2}x\left(t - \frac{1}{t}\right)\right).$$

(By thinking about what happens to the power series when t and $1/t$ are switched, you can see that $J_{-n}(x) = J_n(x)$ for n even and $J_{-n}(x) = -J_n(x)$ for n odd.) $J_n(x)$ is also a solution to *Bessel's equation of order* n,

$$x^2y'' + xy' + (x^2 - n^2)y = 0.$$

The Taylor series expansion for J_0 is (*cf.* Section 5.8 of Boyce & DiPrima):

$$J_0(x) = \sum_{j=0}^{\infty}(-1)^j \frac{1}{2^{2j}} \frac{x^{2j}}{(j!)^2}.$$

(a) MATLAB has a built-in function for $J_n(x)$, called **besselj(n, x)**. Use it to plot $J_0(x)$ for $0 \leq x \leq 10$. (You don't need to graph the function for negative x since J_0 is an even function.) On the same axes, draw the graphs of the 4th order, 10th order, and 20th order Taylor polynomials of J_0. How well do the Taylor polynomials approximate the function?

(b) An interesting equality satisfied by the Bessel functions is

$$J_n'(x) = \frac{1}{2}\left(J_{n-1}(x) - J_{n+1}(x)\right).$$

In particular,

$$J_0'(x) = \frac{1}{2}\left(J_{-1}(x) - J_1(x)\right) = \frac{1}{2}\left(-J_1(x) - J_1(x)\right) = -J_1(x).$$

Thus the maxima and minima of J_0 occur at the zeros of J_1. Using **fzero**, solve the equations $J_0(x) = 0$ and $J_1(x) = 0$ numerically to find the first five zeros and the first five relative extreme points of J_0, starting at $x = 0$. (You will need 10 different starting values—use your graph for guidance.) Compute the differences between successive zeros. Can you make a guess about the periods of the oscillations as $x \to \infty$? (If you did Problem 15 in Problem Set D, you know exactly how the period behaves as $x \to \infty$.)

(c) From the general theory, one knows that Bessel's equation of order n has another linearly independent solution $Y_n(x)$, with a logarithmic singularity as

$x \to 0^+$. MATLAB also has a built-in function, **bessely(n, x)**, for computing $Y_n(x)$. Graph the function $Y_0(x)$ on the interval $0 < x \le 10$. Then compute $\lim_{x \to 0^+} Y_0(x)/\ln x$. Graph the function

$$Y_0(x) - c \ln x$$

on the same interval, where c is the value of the limit you obtained, and observe that the function has a limit as x approaches 0 from the right. Thus the "singular part" of $Y_0(x)$ behaves like $c \ln x$.

7. The first order homogeneous linear differential equation

$$x^2 y' + y = 0$$

has an irregular singular point at $x = 0$ (why?); however, as a first order linear equation it can be solved exactly.

(a) Find the general solution of the equation for $x > 0$ and for $x < 0$. How do the solutions behave as $x \to 0^+$? as $x \to 0^-$? Use MATLAB to graph a solution for $0 < x < 2$ and a solution for $-2 < x < 0$.

(b) Now find (for $x > 0$ and for $x < 0$) the general solution of the Euler equation

$$xy' + ry = 0,$$

which has a regular singular point at $x = 0$ if $r \neq 0$. How do the solutions behave as $x \to 0^+$? as $x \to 0^-$? Use MATLAB to graph solutions to the right and left of 0 for each of the cases $r = 0.5$, $r = 1$, $r = -0.5$, $r = -1$. (Be careful! If you compute a general solution, then you will have to allow complex values of the undetermined constant to get all the real solutions.)

(c) What differences do you notice between the behavior of solutions near a regular singular point and the behavior near an irregular singular point?

8. Consider the Legendre equation

$$(1 - x^2)y'' - 2xy' + \alpha(\alpha + 1)y = 0. \tag{E.2}$$

We assume that $|x| < 1$, because $x = \pm 1$ are regular singular points. We also assume that $\alpha > -1$.

(a) Solve the equation (E.2) with **dsolve** for $\alpha = -1/2$ subject to the initial condition $y(0) = 0$, $y'(0) = 1$. As you can see, the solution is given in terms of unfamiliar special functions, which MATLAB finds difficult to evaluate. Therefore find a series approximation (out to terms in x^9) of the solution to the same equation and plot the result. Compare with a plot of a numerical solution of the same initial value problem.

(b) If $\alpha = n$ is a nonnegative integer, the Legendre equation has a polynomial solution of order n. The *Legendre polynomial* $P_n(x)$ is defined to be the unique polynomial solution of the Legendre equation (with $\alpha = n$) such that $P_n(1) = 1$. Use MATLAB to compute the first six Legendre polynomials $P_0(x), \ldots, P_5(x)$ as follows. First use **dsolve** (without initial conditions) to solve the Legendre equation for $n = 0, \ldots, 5$. Then inspect each solution and choose values for the undetermined constants to get polynomial solutions with $P_n(1) = 1$.

(c) Graph $P_0(x), \ldots, P_5(x)$ on the interval $[-1, 1]$.

(d) In the graph in part (c), you can see that as n increases, the Legendre polynomials are more and more oscillatory on the interval $[-1, 1]$. For x close to zero, the Legendre equation is approximately $y'' + n(n+1)y = 0$. By solving this equation, explain why solutions of the Legendre equation display increasing oscillation (on the interval $[-1, 1]$) as $n \to \infty$.

9. Nonlinear differential equations can sometimes be solved by series methods. Consider, for example,

$$y' = y^2 + x^2, \tag{E.3}$$

an equation, discussed in Example 5.7, whose solution via **dsolve** leads to problematic results. Let's use the initial condition $y(0) = 0$, and look for a solution of the form

$$y(x) = \sum_{i=1}^{\infty} a_i x^i.$$

(We know there is no constant term in the Taylor series expansion of y at $x = 0$ because of the initial condition.)

(a) Explain why a_1 must equal 0.

(b) Now use the scheme from Chapter 11 to compute a Taylor series expansion of the solution to (E.3). Make sure you go far enough to get the 11th degree monomial. What pattern do you see?

(c) In particular, only odd powers of x appear. Show that the solution to the IVP $y' = y^2 + x^2$, $y'(0) = 0$ must be an odd function, that is, $y(-x) = -y(x)$. (*Hint*: Set $w(x) = -y(-x)$ and show that w solves the IVP.)

(d) Plot the 11th degree Taylor polynomial for y, together with a numerical solution of (E.3) obtained using **ode45**, on the interval $-1 < x < 1$. How close are the graphs? Now try plotting on the interval $-1.95 < x < 1.95$. What is happening? Next note that for $x > 0$, we have $y'(x) > x^2$, so $y(x) > x^3/3$. In particular, $y(1) > 1/3$. But for $x > 1$, the right-hand side of (E.3) is greater than $y(x)^2 + 1$; so the solution grows faster than the solution to the IVP $y' = y^2 + 1$, $y(1) = 1/3$, Use **dsolve** to solve that IVP. You get a solution of the form $y_1(x) = \tan(x - C)$. Using the fact that the tangent function blows up when its argument reaches $\pi/2$, find the value v_1 at which y_1 blows up. (It will be approximately 2.2.) Hence the solution of (E.3) must become infinite somewhere between $x = 1$ and $x = v_1$. Is there any way of seeing this from the series solution? Use **ode45** to locate the point P where the solution blows up. Graph the two approximate solutions (that is, the series solution out to degree 11 and the numerical solution) on the interval $(0, P - \epsilon)$, where you will have to pick the small positive number ϵ carefully to get a useful plot. What do you observe?

10. For each of the following functions, compute the Laplace Transform. Then plot the Laplace Transform over the interval $[0, 10]$.

    (a) $\sin t$.

    (b) e^t.

    (c) $\cos t$.

    (d) $t \cos t$.

    (e) $u_1(t) \sin t$.

    On the basis of these graphs, what general conclusions can you draw about the Laplace Transform of a function? In particular, what can you say about the growth or decay of the Laplace Transform of a function? Can you justify your conclusion by looking at the integral formula for the Laplace Transform? (*Hint*: Look at the derivation of $\mathcal{L}(e^{at})$ at the beginning of Chapter 12.)

11. This problem is intended only for those with access to the Professional Version of MATLAB. Suppose $f(t)$ and $g(t)$ are functions with Laplace Transforms $\mathcal{L}(f) = F$ and $\mathcal{L}(g) = G$. Check the following identities with MATLAB. You should begin with the commands

```
>> maple('addtable', 'laplace', 'f(t)', 'F(s)', t, s);
>> maple('addtable', 'laplace', 'g(t)', 'G(s)', t, s);
```

    which load the information that $\mathcal{L}(f) = F$ and $\mathcal{L}(g) = G$ into MATLAB's memory. (These commands will not work in the Student Version.)

    (a) $\mathcal{L}(af + bg) = a\mathcal{L}(f) + b\mathcal{L}(g)$, where a and b are real numbers.

    (b) $\mathcal{L}(u_1(t)f(t-1)) = e^{-s}F(s)$.

    (c) $\mathcal{L}(e^{ct}f(t)) = F(s-c)$.

    (d) $\mathcal{L}(f') = sF(s) - f(0)$.

    (e) $\mathcal{L}(\int_0^t f(t-u)g(u)\,du) = F(s)G(s)$.

    The function $t \to \int_0^t f(t-u)g(u)\,du$ is called the *convolution* of f and g, and is written $f * g$. Thus part (e) says that the Laplace Transform changes convolution into ordinary multiplication.

12. Consider the following IVP:

$$y''+2y'+2y=g(t),\quad y(0)=0,\quad y'(0)=0,\quad g(t)=\begin{cases}\sin t, & \text{if } 0<\pi\\ 0, & \text{if } t\geq\pi.\end{cases}\tag{E.4}$$

 (a) Solve the equation for $0 \leq t \leq \pi$ by using **dsolve** on the IVP

$$y''+2y'+2y=\sin t,\quad y(0)=0,\quad y'(0)=0.$$

 Plot the solution on the interval $0 \leq t \leq \pi$.

 (b) Find numerical values for $y_\pi = y(\pi)$ and $y'_\pi = y'(\pi)$ using the solution from part (a). Then use **dsolve** to solve the IVP

$$y''+y'+2y=0,\quad y(\pi)=y_\pi,\quad y'(\pi)=y'_\pi.$$

 Plot the solution on the interval $\pi \leq t \leq 15$. Combine this graph with your graph from part (a) to show the solution to (E.4) on the entire interval $0 \leq t \leq 15$.

 (c) Now solve the equation using the Laplace Transform method. Plot the solution on $[0, 15]$, and compare it with the solution from part (b).

 (d) Find the general solution of the associated homogeneous equation. What is the asymptotic behavior of those solutions as $t \to \infty$? Knowing that the forcing term of the inhomogeneous equation is zero after $t = \pi$, what can you say about the long-term behavior of the solution of the inhomogeneous equation?

13. Use the Laplace Transform method to solve the following initial value problems. Then graph the solutions on the interval $[0, 15]$. In parts (a) and (b), also graph the function on the right-hand side of the differential equation (the forcing function). In each part, discuss how the solution behaves in response to the forcing function.

 (a) $y''+3y'+2y=h(t),\quad y(0)=1,\quad y'(0)=0,$

$$h(t)=\begin{cases}1, & \text{if } \pi\leq t\leq 10\\ 0, & \text{if } t>10.\end{cases}$$

 (b) $y''+2y'+3y=u_3(t),\quad y(0)=0,\quad y'(0)=1.$

 (c) $y''+2y'+\frac{4}{5}y=g(t),\quad y(0)=0,\quad y'(0)=0,$

$$g(t)=\begin{cases}\cos(t), & \text{if } 0\leq t\leq\pi\\ 0, & \text{if } t>\pi.\end{cases}$$

 (d) $y''+4y=\delta(t-\pi)-\delta(t-2\pi),\quad y(0)=0,\quad y'(0)=0.$

(e) $y'' + 2y' + 3y = \sin t + \delta(t - 3\pi), \quad y(0) = 0, \quad y'(0) = 0.$

14. Use the Laplace Transform method to solve the following initial value problems. See Problem 13 for additional instructions.

(a) $y'' + 4y = \sin t - u_{2\pi}(t)\sin(t - 2\pi), \quad y(0) = 0, \quad y'(0) = 0.$

(b) $y'' + 6y' + 8y = h(t), \quad y(0) = 0, \quad y'(0) = 2, \quad h(t) = \begin{cases} 0, & 0 \le t < 5 \\ 1, & 5 \le t < 10 \\ -1, & t \ge 10. \end{cases}$

(c) $y'' + 4y = \delta(t - 3\pi), \quad y(0) = 1, \quad y'(0) = 0.$

(d) $y'' + y = \delta(t - 2) - \delta(t - 8), \quad y(0) = 0, \quad y'(0) = 0.$

15. Use the Laplace Transform method to solve the following initial value problems of higher order. Instead of cumbersome expressions like **'D(D(D(D(y))))(t)'**, MATLAB allows you to type **'diff(y(t), t$4)'**. You will discover, after taking the Laplace Transform, that MATLAB gives the higher order initial conditions in terms of strange constructs like **@@(D,2)(y)(0)**, which **subs** cannot handle. Thus to set $y''(0) = 1$, for example, you can use the following substitute:

```
>> neweqn = sym(strrep(char(lteqn), ...
     '`@@`(D,2)(y)(0)', '1'))
```

Here we are using MATLAB's command **char** to convert from a symbolic expression to a string, **strrep** to replace part of a string, and then **sym** to convert back from a string to a symbolic expression. The back quotes flag **@@** as a special symbol (which comes from the Maple kernel).

Once you have solved the equation, plot the solution on an appropriate interval ($t > 0$).

(a) $y'''(t) - y''(t) - y'(t) - 2y(t) = \delta(t - 1), \quad y(0) = y'(0) = y''(0) = 0.$

(b) $y^{(4)}(t) + 2y''(t) + y(t) = \cos t, \quad y(0) = 1, \quad y'(0) = y''(0) = y'''(0) = 0.$

(c) $y^{(4)}(t) + 3y''(t) - 4y(t) = u_1(t)\sin(2t), \quad y(0) = y'(0) = y''(0) = y'''(0) = 0.$

(d) $y'''(t) + 4y'(t) = \delta(t - \pi), \quad y(0) = y''(0) = 0, \quad y'(0) = 1.$

Which of these equations has resonance-type behavior?

16. This problem is based on Problem 35 in Section 6.2 of Boyce & DiPrima. Consider Bessel's equation

$$ty'' + y' + ty = 0.$$

We will use the Laplace Transform to find some leading terms of the power series expansion of the Bessel function of order zero that is continuous at the origin. Let $y(t)$ be such a solution of the equation, and suppose $Y(s)$ is its transform.

(a) Show that $Y(s)$ satisfies the equation

$$(1 + s^2)Y'(s) + sY(s) = 0.$$

(b) Solve the equation in part (a) using **dsolve**. (Use $Y(0) = 1$.)

(c) Use **taylor** to expand the solution in part (b) in powers of $1/s$ out to terms of degree 10. (*Hint*: The expansion should be valid as $s \to \infty$; if you ask **taylor** to expand about **inf** you will get a series in powers of $1/s$.)

(d) Apply the Inverse Laplace Transform to part (c) to obtain the power series expansion of y in powers of t. To what degree is this power series valid?

17. This problem illustrates how the choice of method can dramatically affect the time it takes the computer to solve a differential equation. It can also affect the form of the solution. Consider the initial value problem

$$y'' + y' + y = (t + 1)^3 e^{-t} \cos t, \qquad y(0) = 1, \qquad y'(0) = 0. \tag{E.5}$$

This problem could be solved by the methods of undetermined coefficients, variation of parameters, or Laplace Transforms.

(a) Try to solve (E.5) using **dsolve**. What happens?

(b) If you have the Professional Version of MATLAB, you can see what went wrong in part (a) by executing the command

```
>> maple('dsolve', ['D(D(y))(t) + D(y)(t) + ', ...
      'y(t) = (t+1)^3*exp(-t)*cos(t)'])
```

What do you get? MATLAB's Maple kernel has used a solution method that requires solving a really messy algebraic equation, hence the answer comes out in an unusable form.

(c) Now use the Laplace Transform method from Chapter 12 to solve the problem. How long does it take to get a result?

(d) Verify that the expression produced in part (c) really is a solution to the differential equation. Plot the solution over the range $0 \le t \le 15$.

18. In this problem we investigate the effect of a periodic discontinuous forcing function (a square wave) on a second order linear equation with constant coefficients. Consider the initial value problem

$$y'' + y = h(t), \quad y(0) = 0, \; y'(0) = 1. \tag{E.6}$$

The associated homogeneous equation has natural period 2π; the general solution of the homogeneous equation is

$$y(t) = A \cos t + B \sin t.$$

Recall that the phenomenon of resonance occurs when the forcing function $h(t)$ is a linear combination of $\sin t$ and $\cos t$. Does resonance occur when the forcing function is periodic of period 2π but discontinuous?

(a) Using step functions, define a MATLAB function $h(t)$ on the interval $[0, 10\pi]$ whose value is $+1$ on $[0, \pi)$, -1 on $[\pi, 2\pi)$, $+1$ on $[2\pi, 3\pi)$, and so on. Plot the function on the interval $[0, 30]$. It should have the appearance of a square wave.

(b) Use the Laplace Transform method from Chapter 12 to solve equation (E.6) with the function $h(t)$ defined in part (a). Plot the solution together with $h(t)$ on the interval $[0, 30]$. Do you see resonance?

(c) In part (a), we constructed a forcing function $h(t)$ with period 2π. The function $h(t/2)$ has period 4π. Repeat part (b) using the forcing function $h(t/2)$. Do you see resonance?

(d) Repeat part (b) using the forcing function $h(2t)$; this time just plot from 0 to 15. Do you see resonance? What is the period of $h(2t)$?

(e) What can you conclude about the resonance effect for discontinuous forcing functions? Would you expect resonance to occur in equation (E.6) for *any* forcing function of period 2π? (*Hint*: The function $h(2t)$ has period π as well as period 2π.) Can you venture a guess about when resonance occurs for discontinuous periodic forcing functions? You might try some other periodic forcing functions to check your guess.

19. One additional use of the Laplace Transform is to transform a *partial differential equation*, involving both time and space derivatives, into an equation involving only space derivatives. We will illustrate this process with the *heat equation* in one space variable, which models the temperature $u(x, t)$ in a rod as a function of position x and time t. This equation takes the form

$$\frac{\partial u}{\partial t} = k \frac{\partial^2 u}{\partial x^2}, \tag{E.7}$$

where k is a constant depending on the material the rod is made of. For simplicity, we will assume units have been chosen so that $k = 1$. Let's assume our rod has length 1 (again, in suitable units) and that the temperature at the two ends of the rod is kept fixed, say at $u = 0$. Then we have a *boundary value problem* of the form

$$u(0, t) = u(1, t) = 0, \qquad u(x, 0) = u_0(x), \tag{E.8}$$

where u_0 is the initial temperature distribution in the rod at $t = 0$. Note that u_0 should satisfy $u_0(0) = u_0(1) = 0$, for compatibility with (E.8).

We can solve this problem as follows. Take the Laplace Transform $U(x, s)$ of $u(x, t)$ in t, thinking of x as an extra parameter. Then (E.7) and (E.8) (with $k = 1$) become a boundary value problem for an ordinary differential equation for U, with x as the independent variable and s as an extra parameter:

$$sU(x, s) - u_0(x) = \frac{d^2U(x, s)}{dx^2}, \qquad U(0, s) = U(1, s) = 0. \tag{E.9}$$

We can solve (E.9), say with **dsolve**, and then take the Inverse Laplace Transform to recover the function u.

(a) Carry out this process to solve (E.7) and (E.8) (with $k = 1$) when the initial temperature is $u_0(x) = \sin(\pi x)$. You should get a nice simple formula for the answer; check that it indeed satisfies (E.7).

(b) Repeat part (a) with $u_0(x) = \sin(n\pi x)$, $n = 2, 3, 4$. (You can do this with a loop.) You should see a general pattern! What is the solution for general n? Check it! (Note: It doesn't work simply to declare n to be a symbolic variable and to try to do the general case all at once, because MATLAB doesn't know you want to constrain n to be an integer, and if it is *not* an integer, the boundary condition won't be satisfied.)

(c) Attempt to repeat the process with $u_0(x) = x(1 - x)$. You should find that MATLAB solves the ODE with no difficulty, but cannot compute the Inverse Laplace Transform to recover u. (This is symptomatic of the fact that there is no formula for u in terms of elementary functions in this case.)

(d) To deal with the problem in (c), there is another method. Attempt to write $u_0(x) = x(1 - x)$ as a linear combination of $\sin(n\pi x)$ for different values of n. This gives what is called the *Fourier expansion* of the function. Since you know the solution to (E.7) and (E.8) in this case, and since the equation is linear, that enables you to write down the solution to the problem as an infinite series, using what is sometimes called the *principle of superposition.* To find the coefficients in the Fourier expansion, suppose you have a formal expansion

$$u_0(x) = x(1 - x) = \sum_{n=1}^{\infty} c_n \sin(n\pi x), \quad 0 \le x \le 1. \tag{E.10}$$

The coefficients c_n are computed as follows. Multiply both sides of (E.10) by $\sin(m\pi x)$ for fixed m, and integrate from 0 to 1. You can check using **int** that

$$\int_0^1 \sin^2(n\pi x)\,dx = \frac{1}{2}, \qquad \int_0^1 \sin(n\pi x)\sin(m\pi x)\,dx = 0, \quad n \ne m.$$

Thus one gets

$$\begin{aligned}\int_0^1 \left(\sum_{n=1}^{\infty} c_n \sin(n\pi x)\right)\sin(m\pi x)\,dx \\ &= \sum_{n=1}^{\infty} c_n \int_0^1 \sin(n\pi x)\sin(m\pi x)\,dx \\ &= \sum_{n=1}^{\infty} c_n \left(\tfrac{1}{2} \text{ if } m = n, 0 \text{ otherwise}\right) = \frac{c_m}{2},\end{aligned}$$

and so

$$c_m = 2\int_0^1 x(1 - x)\sin(m\pi x)\,dx.$$

Compute the coefficients c_m for $1 \leq m \leq 10$ and plot

$$u_0(x) \quad \text{and} \quad \sum_{n=1}^{10} c_n \sin(n\pi x)$$

on the same axes to see the agreement. Then find the corresponding solution of (E.7) and (E.8) and plot it as a function of x and t. (For this you can use the function **ezmesh**—see the online help for the syntax. It should suffice to take $0 \leq x,\, t \leq 1$.) How does the solution match your intuition that temperature fluctuations in the rod at $t = 0$ should dissipate with time?

20. Consider the mass-damper system with dead zone as depicted in Figure E.1. The motion of the system is described by the following combination of equations

$$\begin{cases} my'' + 2cy' + k(y+b) = F(t), & y < -b \\ my'' + 2cy' = F(t), & -b \leq y \leq b \\ my'' + 2cy' + k(y-b) = F(t), & y > b. \end{cases} \tag{E.11}$$

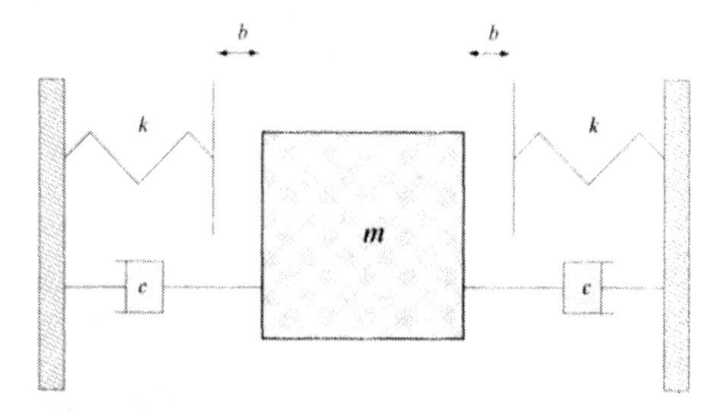

Figure E.1: A Mass-Damper System with a Dead Zone

(a) Rewrite equation (E.11) as a single equation using the **heaviside** function.

(b) Consider a free system, $F \equiv 0$, and let $m = 10$kg, $c = 50$Ns/m and $k = 150$N/m. Suppose the motion starts at $y(0) = 0$ with initial velocity $v_0 = y'(0) = 20$m/s. Finally take $b = 0.5$m. Find a numerical solution with **ode45** and graph it on the interval $0 \leq t \leq 15$.

(c) Now by varying the friction constant c, the initial velocity v_0 and the dead zone size b, exhibit at least two additional types of behavior for the solution function.

(d) Repeat parts (b) and (c) in the presence of a forcing function

$$F(t) = 20\cos(12t).$$

21. Consider the "tunable" RLC circuit illustrated in Figure E.2. It consists of a variable resistor with resistance R, an inductance L, and a capacitor with capacitance C. The circuit can be tuned by adjusting R. If an input voltage $f(t)$ is applied as shown, the voltage v across the resistor satisfies the differential equation

$$v'' + \frac{R}{L}v' + \frac{1}{LC}v = \frac{R}{L}f'. \tag{E.12}$$

Assume that in suitable units, $LC = 1$.

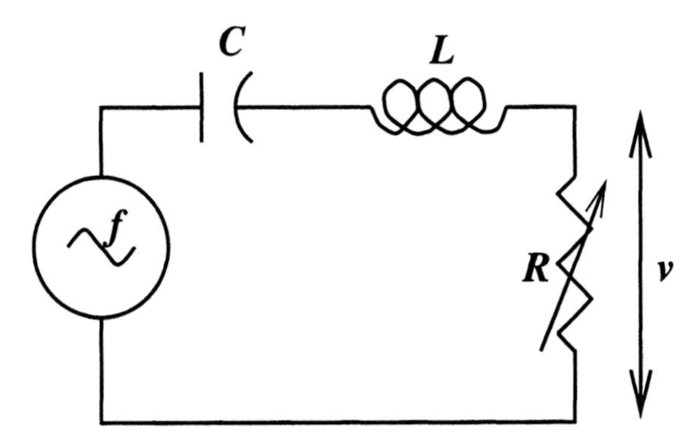

Figure E.2: An LRC Circuit with an Input Pulse

(a) Build a Simulink model for the equation (E.12), that takes the parameter $r = \frac{R}{L}$ from a Constant or From Workspace block and the forcing term $f'(t)$ from a Pulse Generator block (that produces a square wave). You can use the default parameters $v(0) = v'(0) = 0$ for the initial conditions and the default parameters for the square wave (amplitude 1, period 2, and pulse width 1). (Note that this corresponds to an input voltage $f(t)$ that is the *integral* of this square wave.) Let t run from 0 to 20 in the simulation. Experiment with "tuning" the circuit by varying the parameter $r = \frac{R}{L}$. Try $r = 0.1,\ 1,\ 2,\ 3,\ 10$. You should observe very different behavior of the output voltage $v(t)$, depending on the value of r. What happens? Can you understand why? (*Hint*: Solve (E.12) with **dsolve** in the homogeneous case $f \equiv 0$. How do the solutions behave?)

(b) Solve the equation (E.12), with $r = \frac{R}{L}$ declared to be a symbolic variable and initial conditions $v(0) = v'(0) = 0$, using **dsolve**, in the case where the input voltage $f(t) = u_1(t)$ is a step function. (Note that this means that $f'(t)$, which is what actually appears in the equation, is the Dirac delta function $\delta_1(t)$.) Plot the solution for $r = 0.1,\ 1,\ 2,\ 3,\ 10$. Do you observe the same phenomena as in part (a)?

Chapter 13

Higher Order Equations and Systems of First Order Equations

First and second order differential equations arise naturally in many applications. For example, first order differential equations occur in models of population growth and radioactive decay, and second order equations in the study of the motion of a falling body or the motion of a pendulum. There are several techniques for solving special classes of first and second order equations. These techniques produce symbolic solutions, expressed by a formula. For equations that cannot be solved by any of these techniques, we use numerical, geometric, or qualitative methods to investigate solutions.

Equations of higher order, and systems of first order equations, also arise naturally. For example, third order equations come up in fluid dynamics; fourth order equations in elasticity; and systems of first order equations in the study of spring–mass systems with several springs and masses. For general higher order equations and systems, there are hardly any techniques for obtaining explicit formula solutions, and numerical, geometric, or qualitative techniques must be used. Nevertheless, for the special class of constant coefficient linear equations and first order linear systems, there are techniques for producing explicit solutions.

In this chapter, we consider higher order equations and first order systems. We outline the basic theory in the linear case. We show how to solve linear first order systems, first by using MATLAB to compute eigenpairs of the coefficient matrix and then by using **`dsolve`**. Then we discuss the plotting of phase portraits, both using exact solutions and, as is generally necessary for nonlinear systems, using numerical solutions. In particular, at the end of this chapter we show how to use **`ode45`** to solve systems of equations numerically.

13.1 Higher Order Linear Equations

Consider the linear equation

$$y^{(n)} + p_1(t)y^{(n-1)} + \cdots + p_n(t)y = g(t), \tag{13.1}$$

and assume that the coefficient functions are continuous on the interval $a < t < b$. Then the following are true:

- If $y_1(t), y_2(t), \ldots, y_n(t)$ are n linearly independent solutions of the corresponding homogeneous equation,

$$y^{(n)} + p_1(t)y^{(n-1)} + \cdots + p_n(t)y = 0, \tag{13.2}$$

  and if y_p is any particular solution of the inhomogeneous equation (13.1), then the general solutions of the equations (13.1), (13.2) are, respectively,

$$y(t) = \sum_{i=1}^{n} C_i y_i(t) + y_p(t)$$

  and

$$y(t) = \sum_{i=1}^{n} C_i y_i(t).$$

- Given a point $t_0 \in (a, b)$ and initial values $y(t_0) = y_0, y'(t_0) = y_0', \ldots, y^{(n-1)}(t_0) = y_0^{(n-1)}$, there is a unique solution to equation (13.1) satisfying the initial data; the solution has n continuous derivatives and exists on the entire interval (a, b).

- A set of n solutions $y_1(t), y_2(t), \ldots, y_n(t)$ of (13.2) is linearly independent if and only if the Wronskian

$$W(y_1, y_2, \ldots, y_n) = \det \begin{pmatrix} y_1 & y_2 & \cdots & y_n \\ y_1' & y_2' & \cdots & y_n' \\ \vdots & \vdots & \ddots & \vdots \\ y_1^{(n-1)} & y_2^{(n-1)} & \cdots & y_n^{(n-1)} \end{pmatrix}$$

  is nonzero at some point in (a, b); a linearly independent collection $y_1, y_2, \ldots, y_n$ is called a *fundamental set* of solutions of (13.2).

We see that the general theory of higher order linear equations parallels that of first and second order linear equations. We note, however, that solutions may be difficult to find. In practice, we can find fundamental sets of solutions only in special cases such as constant coefficient and Euler equations. In the latter cases, solutions can be found by considering trial solutions e^{rt} and t^r, as in the second order case. There is a difficulty even for these equations if n is large, since it is difficult to compute the roots of the characteristic polynomial. We can, however, use MATLAB to find good approximations to the roots.

The **dsolve** command can solve many homogeneous constant coefficient and Euler equations. You can confirm this by solving several linear equations of higher order from your text.

13.2 Systems of First Order Equations

Consider the general system of first order equations,

$$\begin{aligned} x_1' &= F_1(t, x_1, x_2, \dots, x_n) \\ x_2' &= F_2(t, x_1, x_2, \dots, x_n) \\ &\vdots \\ x_n' &= F_n(t, x_1, x_2, \dots, x_n). \end{aligned} \tag{13.3}$$

Such systems arise in the study of spring–mass systems with many springs and masses, and in many other applications. Systems of first order equations also arise in the study of higher order equations. A single higher order differential equation

$$y^{(n)} = F(t, y, y', \dots, y^{(n-1)})$$

can be converted into a system of first order equations by setting

$$x_1 = y, x_2 = y', \dots, x_n = y^{(n-1)}.$$

The resulting system is

$$\begin{aligned} x_1' &= x_2 \\ x_2' &= x_3 \\ &\vdots \\ x_{n-1}' &= x_n \\ x_n' &= F(t, x_1, x_2, \dots, x_n). \end{aligned}$$

It is not surprising that the problem of analyzing the general system (13.3) is quite formidable. Generally, the techniques we can bring to bear will be numerical, geometric, or qualitative. We shall discuss the application of these techniques to nonlinear equations at the end of this chapter and in Chapter 14. For now, we merely observe that we can expect to make some progress toward a formula solution for *linear systems*.

13.2.1 Linear First Order Systems

Consider a linear system

$$\begin{aligned} x_1' &= a_{11}(t)x_1 + \cdots + a_{1n}(t)x_n + b_1(t) \\ x_2' &= a_{21}(t)x_1 + \cdots + a_{2n}(t)x_n + b_2(t) \\ &\vdots \\ x_n' &= a_{n1}(t)x_1 + \cdots + a_{nn}(t)x_n + b_n(t). \end{aligned}$$

We can write this system in matrix notation:

$$\mathbf{X}' = \mathbf{A}(t)\mathbf{X} + \mathbf{B}(t),$$

where

$$\mathbf{X}(t) = \begin{pmatrix} x_1(t) \\ x_2(t) \\ \vdots \\ x_n(t) \end{pmatrix}, \qquad \mathbf{B}(t) = \begin{pmatrix} b_1(t) \\ b_2(t) \\ \vdots \\ b_n(t) \end{pmatrix},$$

and

$$\mathbf{A}(t) = \begin{pmatrix} a_{11}(t) & a_{12}(t) & \dots & a_{1n}(t) \\ \vdots & \vdots & \ddots & \vdots \\ a_{n1}(t) & a_{n2}(t) & \dots & a_{nn}(t) \end{pmatrix}.$$

There is a general theory for linear systems, which is completely parallel to the theory for a single linear equation. If $\mathbf{X}^{(1)}, \mathbf{X}^{(2)}, \dots, \mathbf{X}^{(n)}$ is a family of linearly independent solutions, *i.e.*, a *fundamental set* of solutions, of the homogeneous equation (with $\mathbf{B} = \mathbf{0}$), and if $\mathbf{X}_p$ is a particular solution of the inhomogeneous equation, then the general solutions of these equations are, respectively,

$$\mathbf{X} = \sum_{i=1}^{n} C_i \mathbf{X}^{(i)}$$

and

$$\mathbf{X} = \sum_{i=1}^{n} C_i \mathbf{X}^{(i)} + \mathbf{X}_p.$$

The test for linear independence of n solutions is

$$\det(\mathbf{X}^{(1)} \dots \mathbf{X}^{(n)}) \neq 0.$$

For linear systems, the constant coefficient case is the easiest to handle, just as it was for single linear equations. Consider the homogeneous linear system

$$\mathbf{X}' = \mathbf{A}\mathbf{X}, \tag{13.4}$$

where

$$\mathbf{A} = \begin{pmatrix} a_{11} & a_{12} & \dots & a_{1n} \\ \vdots & \vdots & \ddots & \vdots \\ a_{n1} & a_{n2} & \dots & a_{nn} \end{pmatrix}$$

is an $n \times n$ matrix of real constants. We seek nonzero solutions of the form $\mathbf{X}(t) = \boldsymbol{\xi} e^{rt}$; note the similarity to the case of a single linear equation. We find that $\mathbf{X}(t)$ is a solution of (13.4) if and only if

$$\mathbf{A}\boldsymbol{\xi} = r\boldsymbol{\xi},$$

i.e., if and only if r is an *eigenvalue* of $\mathbf{A}$ and $\boldsymbol{\xi}$ is a corresponding *eigenvector*. We call the pair $r, \boldsymbol{\xi}$ an *eigenpair*. In order to explain how to construct a fundamental set of solutions, we have to consider three separate cases.

Case I. $\mathbf{A}$ has distinct real eigenvalues.

Let $r_1, r_2, \ldots, r_n$ be the distinct eigenvalues of $\mathbf{A}$, and let $\boldsymbol{\xi}^{(1)}, \boldsymbol{\xi}^{(2)}, \ldots, \boldsymbol{\xi}^{(n)}$ be corresponding eigenvectors. Then

$$\mathbf{X}^{(1)}(t) = \boldsymbol{\xi}^{(1)} e^{r_1 t}, \ldots, \mathbf{X}^{(n)}(t) = \boldsymbol{\xi}^{(n)} e^{r_n t}$$

is a fundamental set of solutions for (13.4), and

$$\mathbf{X}(t) = c_1 \boldsymbol{\xi}^{(1)} e^{r_1 t} + \cdots + c_n \boldsymbol{\xi}^{(n)} e^{r_n t}$$

is the general solution.

Case II. $\mathbf{A}$ has distinct eigenvalues, some of which are complex.

Since $\mathbf{A}$ is real, if $r, \boldsymbol{\xi}$ is a complex eigenpair, then the complex conjugate $\bar{r}, \bar{\boldsymbol{\xi}}$ is also an eigenpair. Thus the corresponding solutions

$$\mathbf{X}^{(1)}(t) = \boldsymbol{\xi} e^{rt}, \ \mathbf{X}^{(2)}(t) = \bar{\boldsymbol{\xi}} e^{\bar{r}t}$$

are conjugate. Therefore, we can find two real solutions of (13.4) corresponding to the pair r, $\bar{r}$ by taking the real and imaginary parts of $\mathbf{X}^{(1)}(t)$ or $\mathbf{X}^{(2)}(t)$. Writing $\boldsymbol{\xi} = \mathbf{a} + i\mathbf{b}$, where $\mathbf{a}$ and $\mathbf{b}$ are real, and $r = \lambda + i\mu$, where λ and μ are real, we have

$$\begin{aligned} \mathbf{X}^{(1)}(t) &= (\mathbf{a} + i\mathbf{b}) e^{(\lambda + i\mu)t} \\ &= (\mathbf{a} + i\mathbf{b}) e^{\lambda t} (\cos \mu t + i \sin \mu t) \\ &= e^{\lambda t} (\mathbf{a} \cos \mu t - \mathbf{b} \sin \mu t) + i e^{\lambda t} (\mathbf{a} \sin \mu t + \mathbf{b} \cos \mu t). \end{aligned}$$

Thus the vector functions

$$\mathbf{u}(t) = e^{\lambda t} (\mathbf{a} \cos \mu t - \mathbf{b} \sin \mu t)$$

$$\mathbf{v}(t) = e^{\lambda t} (\mathbf{a} \sin \mu t + \mathbf{b} \cos \mu t)$$

are real solutions to (13.4).

To keep the discussion simple, assume that $r_1 = \lambda + i\mu$, $r_2 = \lambda - i\mu$ are complex, and that $r_3, \ldots, r_n$ are real and distinct. Let the corresponding eigenvectors be $\boldsymbol{\xi}^{(1)} = \mathbf{a} + i\mathbf{b}, \boldsymbol{\xi}^{(2)} = \mathbf{a} - i\mathbf{b}, \boldsymbol{\xi}^{(3)}, \ldots, \boldsymbol{\xi}^{(n)}$. Then

$$\mathbf{u}(t), \mathbf{v}(t), \boldsymbol{\xi}^{(3)} e^{r_3 t}, \ldots, \boldsymbol{\xi}^{(n)} e^{r_n t}$$

is a fundamental set of solutions for (13.4). The general situation should now be clear.

Case III. $\mathbf{A}$ has repeated eigenvalues.

We restrict our discussion to the case $n = 2$. Suppose $r = \rho$ is an eigenvalue of $\mathbf{A}$ of multiplicity 2, meaning that ρ is a double root of the characteristic polynomial $\det(\mathbf{A} - r\mathbf{I})$. There are still two possibilities. If we can find two linearly independent eigenvectors $\boldsymbol{\xi}^{(1)}$ and $\boldsymbol{\xi}^{(2)}$ corresponding to ρ, then the solutions $\mathbf{X}^{(1)}(t) = \boldsymbol{\xi}^{(1)} e^{\rho t}$ and $\mathbf{X}^{(2)}(t) = \boldsymbol{\xi}^{(2)} e^{\rho t}$

form a fundamental set. The other possibility is that there is only one (linearly independent) eigenvector $\boldsymbol{\xi}$ with eigenvalue ρ. Then one solution of (13.4) is given by

$$\mathbf{x}^{(1)}(t) = \boldsymbol{\xi} e^{\rho t},$$

and a second by

$$\mathbf{x}^{(2)}(t) = \boldsymbol{\xi} t e^{\rho t} + \boldsymbol{\eta} e^{\rho t},$$

where $\boldsymbol{\eta}$ satisfies

$$(\mathbf{A} - \rho \mathbf{I})\boldsymbol{\eta} = \boldsymbol{\xi}.$$

The vector $\boldsymbol{\eta}$ is called a *generalized eigenvector* of $\mathbf{A}$ for the eigenvalue ρ.

13.2.2 Using MATLAB to Find Eigenpairs

We now show how to use MATLAB to find the eigenpairs of a matrix, and thus to find a fundamental set of solutions and the general solution of a system of linear differential equations with constant coefficients.

Example 13.1 Consider the system

$$\mathbf{x}' = \begin{pmatrix} 3 & -2 & 0 \\ 2 & -2 & 0 \\ 0 & 1 & 1 \end{pmatrix} \mathbf{x}.$$

We enter the coefficient matrix in MATLAB as follows:

```
>> A = [3 -2 0; 2 -2 0; 0 1 1]
A =
     3    -2     0
     2    -2     0
     0     1     1
```

We find the eigenpairs of **A** with the command:

```
>> [xi, R] = eig(sym(A))
xi =
[ -1,  0,  2]
[ -2,  0,  1]
[  1,  1,  1]

R =
[ -1,  0,  0]
[  0,  1,  0]
[  0,  0,  2]
```

The second output matrix **R** contains the eigenvalues -1, 1, and 2 on the diagonal (and zeros off the diagonal). For each eigenvalue, the corresponding column of the first output matrix **xi** is an eigenvector. Note that if you run these commands yourself, your output may list the eigenvalues in a different order, but corresponding eigenvalues and eigenvectors will always be in the same column. In this example, the eigenpairs are:

$$r_1 = -1,\ \boldsymbol{\xi}^{(1)} = \begin{pmatrix} -1 \\ -2 \\ 1 \end{pmatrix},$$

$$r_2 = 1,\ \boldsymbol{\xi}^{(2)} = \begin{pmatrix} 0 \\ 0 \\ 1 \end{pmatrix},$$

$$r_3 = 2,\ \boldsymbol{\xi}^{(3)} = \begin{pmatrix} 2 \\ 1 \\ 1 \end{pmatrix}.$$

Thus a fundamental set of solutions is

$$\mathbf{x}^{(1)}(t) = \begin{pmatrix} -1 \\ -2 \\ 1 \end{pmatrix} e^{-t}, \quad \mathbf{x}^{(2)}(t) = \begin{pmatrix} 0 \\ 0 \\ 1 \end{pmatrix} e^{t}, \quad \mathbf{x}^{(3)}(t) = \begin{pmatrix} 2 \\ 1 \\ 1 \end{pmatrix} e^{2t},$$

and the general solution is

$$\mathbf{x}(t) = c_1 \begin{pmatrix} -1 \\ -2 \\ 1 \end{pmatrix} e^{-t} + c_2 \begin{pmatrix} 0 \\ 0 \\ 1 \end{pmatrix} e^{t} + c_3 \begin{pmatrix} 2 \\ 1 \\ 1 \end{pmatrix} e^{2t}.$$

Suppose we are seeking the specific solution with initial value

$$\mathbf{x}(0) = \begin{pmatrix} 3 \\ 5 \\ 0 \end{pmatrix}.$$

Then the constants c_1, c_2, and c_3 must satisfy

$$c_1 \begin{pmatrix} -1 \\ -2 \\ 1 \end{pmatrix} + c_2 \begin{pmatrix} 0 \\ 0 \\ 1 \end{pmatrix} + c_3 \begin{pmatrix} 2 \\ 1 \\ 1 \end{pmatrix} = \begin{pmatrix} -1 & 0 & 2 \\ -2 & 0 & 1 \\ 1 & 1 & 1 \end{pmatrix} \begin{pmatrix} c_1 \\ c_2 \\ c_3 \end{pmatrix} = \begin{pmatrix} 3 \\ 5 \\ 0 \end{pmatrix}.$$

Notice that the matrix in the above equation is simply **xi**. To solve this linear system, we let **b** be the column vector on the right-hand side and "divide" **xi** into **b** with the MATLAB left division operator, which is written as a backslash:

```
>> b = [3; 5; 0];
>> c = xi\b
[ -7/3]
[    2]
[  1/3]
```

Thus the solution of the linear system is $c_1 = -7/3$, $c_2 = 2$, $c_3 = 1/3$, and the solution to our initial value problem is

$$\mathbf{x}(t) = -\frac{7}{3}\begin{pmatrix} -1 \\ -2 \\ 1 \end{pmatrix} e^{-t} + 2\begin{pmatrix} 0 \\ 0 \\ 1 \end{pmatrix} e^{t} + \frac{1}{3}\begin{pmatrix} 2 \\ 1 \\ 1 \end{pmatrix} e^{2t}.$$

Example 13.2 Consider the system

$$\mathbf{x}' = \begin{pmatrix} 3 & -2 \\ 4 & -1 \end{pmatrix} \mathbf{x} \qquad \text{(Boyce \& DiPrima, Sect. 7.6, Prob. 1).}$$

We solve the corresponding eigensystem as follows:

```
>> A = [3 -2; 4 -1];
>> [xi, R] = eig(sym(A))
xi =
[    1,    1]
[ 1-i, 1+i]

R =
[ 1+2*i,       0]
[       0, 1-2*i]
```

So the eigenpairs are

$$1 + 2i, \begin{pmatrix} 1 \\ 1 - i \end{pmatrix}$$

and

$$1 - 2i, \begin{pmatrix} 1 \\ 1 + i \end{pmatrix}.$$

Note that the second pair is the complex conjugate of the first. Thus a complex fundamental set is

$$\mathbf{X}^{(1)}(t) = \begin{pmatrix} 1 \\ 1 - i \end{pmatrix} e^{(1+2i)t}, \quad \mathbf{X}^{(2)}(t) = \begin{pmatrix} 1 \\ 1 + i \end{pmatrix} e^{(1-2i)t}.$$

The real and imaginary parts $\mathbf{u}(t)$ and $\mathbf{v}(t)$ of the solution $\mathbf{X}^{(1)}(t)$ can be extracted by hand to produce a real set of fundamental solutions:

$$\mathbf{u}(t) = e^t \left[\begin{pmatrix} 1 \\ 1 \end{pmatrix} \cos 2t + \begin{pmatrix} 0 \\ 1 \end{pmatrix} \sin 2t\right]$$

$$\mathbf{v}(t) = e^t \left[-\begin{pmatrix} 0 \\ 1 \end{pmatrix} \cos 2t + \begin{pmatrix} 1 \\ 1 \end{pmatrix} \sin 2t\right].$$

Example 13.3 Consider the system

$$\mathbf{x}' = \begin{pmatrix} 3 & -4 \\ 1 & -1 \end{pmatrix} \mathbf{x} \qquad \text{(Boyce \& DiPrima, Sect. 7.8, Prob. 1).}$$

We solve the corresponding eigensystem as follows:

```
>> A = [3 -4; 1 -1];
>> [xi, R] = eig(sym(A))
xi =
[ 2]
[ 1]

R =
[ 1, 0]
[ 0, 1]
```

The number 1 is listed twice as an eigenvalue, but only one eigenvector is listed. This is MATLAB's way of reporting that 1 is an eigenvalue of multiplicity 2 and that $\begin{pmatrix} 2 \\ 1 \end{pmatrix}$ is the only corresponding eigenvector. Let $\rho = 1$ and $\boldsymbol{\xi} = \begin{pmatrix} 2 \\ 1 \end{pmatrix}$. Then

$$\mathbf{X}^{(1)}(t) = \boldsymbol{\xi} e^{\rho t} = \begin{pmatrix} 2 \\ 1 \end{pmatrix} e^t$$

is one solution. To find a second solution we solve

$$(\mathbf{A} - \rho \mathbf{I})\boldsymbol{\eta} = \begin{pmatrix} 2 & -4 \\ 1 & -2 \end{pmatrix} \boldsymbol{\eta} = \boldsymbol{\xi} = \begin{pmatrix} 2 \\ 1 \end{pmatrix}.$$

We enter the coefficient matrix as **M** and use the previous definition of **xi** to solve for **eta**:

```
>> M = [2 -4; 1 -2];
>> eta = M\xi
eta =
[ 1]
[ 0]
```

(This system actually has an infinite number of solutions; MATLAB returns one of them.) Thus a second solution is

$$\mathbf{X}^{(2)}(t) = \boldsymbol{\xi} t e^{\rho t} + \boldsymbol{\eta} e^{\rho t} = \begin{pmatrix} 2 \\ 1 \end{pmatrix} t e^t + \begin{pmatrix} 1 \\ 0 \end{pmatrix} e^t.$$

The general solution is

$$\mathbf{X}(t) = c_1 \begin{pmatrix} 2 \\ 1 \end{pmatrix} e^t + c_2 \left[\begin{pmatrix} 2 \\ 1 \end{pmatrix} t e^t + \begin{pmatrix} 1 \\ 0 \end{pmatrix} e^t \right].$$

We have shown how to use MATLAB to find eigenpairs, and thus solutions, for linear systems of differential equations with constant coefficients. The solutions to systems of linear equations can also be found with a direct application of **dsolve**.

Example 13.4 Consider the system $x' = y$, $y' = -x$. To find its general solution, type:

```
>> [x, y] = dsolve('Dx = y', 'Dy = -x', 't')
x =
cos(t)*C1+sin(t)*C2
y =
-sin(t)*C1+cos(t)*C2
```

To solve the same system with initial conditions $x(0) = 1$, $y(0) = 0$, you can type:

```
>> [x, y] = dsolve('Dx = y', 'Dy = -x', 'x(0) = 1', ...
           'y(0) = 0', 't')
```

Both methods for solving linear systems are useful. Solving in terms of the eigenpairs of the coefficient matrix is somewhat involved, but yields a simple and useful formula for the solution. The eigenpairs contain valuable information about the solution, which we will exploit further in Problem Set F. On the other hand, it is simpler to use **dsolve**, and for many purposes the solutions generated by **dsolve** are completely satisfactory.

13.3 Phase Portraits

A solution of a 2×2 linear system is a pair of functions $x(t)$, $y(t)$. The plot of a solution as a function of t would be a curve in (t, x, y)-space. Generally, such 3-dimensional plots are too complicated to be illuminating. Therefore, we project the curve from 3-dimensional space into the (x, y)-plane. This plane is called the *phase plane*, and the resulting curve in the phase plane is called a *trajectory*. A trajectory is an example of a *parametrized curve* and is drawn by plotting $(x(t), y(t))$ as the parameter t varies. A plot of a family of trajectories is called a *phase portrait* of the system.

In this section, we describe how to use MATLAB: (1) to solve an initial value problem consisting of a linear system and an initial condition, and plot the corresponding trajectory; (2) to solve a linear system using **dsolve** and draw a phase portrait; and (3) to solve a system using **ode45** and draw a phase portrait.

13.3.1 Plotting a Single Trajectory

Suppose we want to plot the trajectory of the initial value problem

$$x' = -3x + 2y, \quad y' = -x; \qquad x(0) = 1, \quad y(0) = 0.$$

First we enter the initial value problem, solve it, and define MATLAB functions **xf(t)** and **yf(t)** using the formulas given by **dsolve**.

```
>> ivp = 'Dx = -3*x + 2*y, Dy = -x, x(0) = 1, y(0) = 0';
>> [x, y] = dsolve(ivp, 't');
>> xf = @(t) eval(vectorize(x));
>> yf = @(t) eval(vectorize(y));
```

Now the functions **xf** and **yf** can be evaluated at a particular time t, or at a particular sequence or vector of times. To plot the trajectory on a given interval, say $-0.3 \le t \le 5$, we type the following:

```
>> t = -0.3:0.1:5;
>> plot(xf(t), yf(t))
>> xlabel 'x'
>> ylabel 'y'
```

The result is shown in Figure 13.1. To use the sequence of commands above for plotting the solution of a different initial value problem, all you have to do is modify the definitions of **ivp** and **t** appropriately.

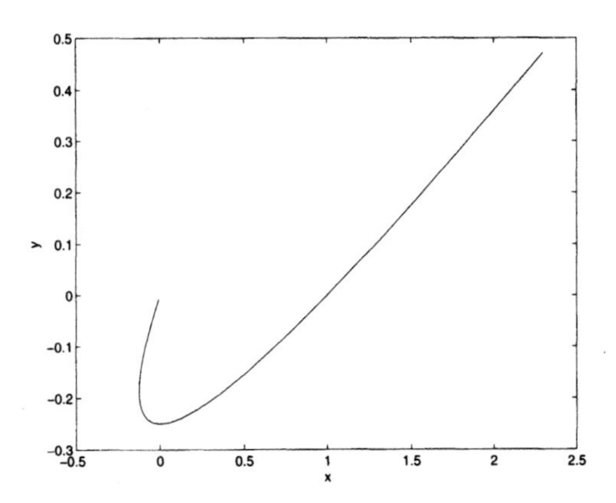

Figure 13.1: A Single Trajectory

13.3.2 Plotting Several Trajectories

Consider now the linear system

$$x' = x - 2y, \quad y' = -x.$$

We want to plot trajectories of this system with various initial conditions $x(0) = a$, $y(0) = b$. We first define functions **xf(t, a, b)** and **yf(t, a, b)** that solve this system with initial conditions $x(0) = a$, $y(0) = b$.

```
>> ivp = 'Dx = x - 2*y, Dy = -x, x(0) = a, y(0) = b';
>> [x, y] = dsolve(ivp, 't');
>> xf = @(t, a, b) eval(vectorize(x));
>> yf = @(t, a, b) eval(vectorize(y));
```

The functions **xf** and **yf** can be evaluated for a particular choice of initial conditions $x(0) = a$, $y(0) = b$, at a particular value of t.

Next we plot a phase portrait of the system.

```
>> figure; hold on
>> t = -3:0.1:3;
>> for a = -2:2
       for b = -2:2
          plot(xf(t, a, b), yf(t, a, b))
       end
   end
>> hold off
>> axis([-20 20 -15 15])
>> xlabel 'x'
>> ylabel 'y'
```

We chose a rectangular grid of 25 initial conditions, with a ranging from -2 to 2 and b ranging from -2 to 2 (in integer increments). We let t range from -3 to 3 in increments of 0.1 and used **axis** to show only the region $-20 \leq x \leq 20$, $-15 \leq y \leq 15$. The result is shown in Figure 13.2.

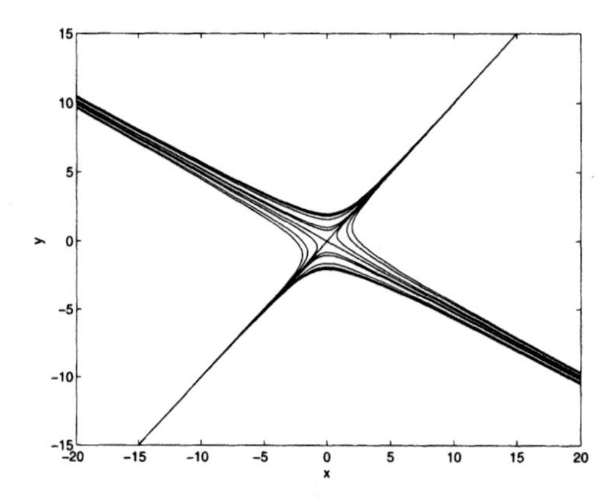

Figure 13.2: A Family of Curves

You should be able to plot the phase portraits for any 2×2 linear system with constant coefficients using the procedure above, replacing the differential equations in the definition of **ivp**, and choosing, by trial and error, an appropriate time interval and appropriate initial conditions to use in defining **t**, **a**, and **b**. By combining the phase portrait with a plot of the vector field (see Chapter 14), you should be able to determine the direction of increasing t on the curves. Also, in each case you should be able to decide whether the origin is a *sink* (all solutions approach the origin as t increases), *source* (all solutions move away from the origin), *saddle point* (solutions approach the origin along one direction, then move away from the origin in another direction), or *center* (solutions follow closed curves around the origin).

The procedure for plotting a phase portrait also works if you wish to vary only one parameter in the initial data. For example, consider the initial value problem

$$x' = x + 2y, \quad y' = -x; \qquad x(0) = a, \quad y(0) = 0$$

for various values of a. The commands

```
>> ivp = 'Dx = x + 2*y, Dy = -x, x(0) = a, y(0) = 0';
>> [x, y] = dsolve(ivp, 't');
>> xf = @(t, a) eval(vectorize(x));
>> yf = @(t, a) eval(vectorize(y));
>> figure; hold on
>> t = -10:0.1:10;
>> for a = -4:4
     plot(xf(t, a), yf(t, a))
   end
>> hold off
>> axis([-15 15 -10 10])
>> xlabel 'x'
>> ylabel 'y'
```

yield the nine curves shown in Figure 13.3. (Count the curves! One of the solutions is the equilibrium solution $x(t) = 0$, $y(t) = 0$.)

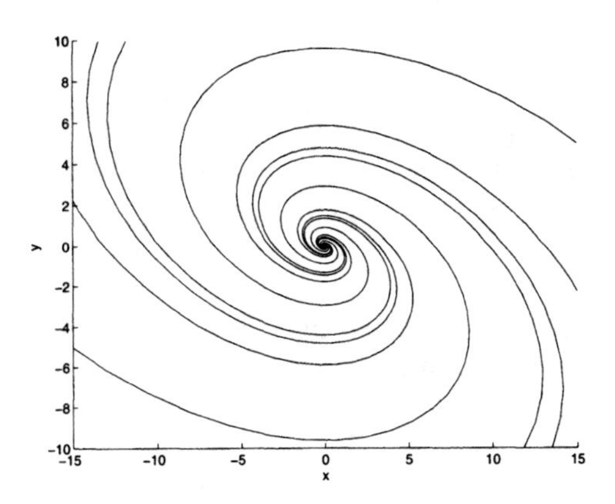

Figure 13.3: Another Family of Curves

13.3.3 Numerical Solutions of First Order Systems

The symbolic solver **dsolve** should be able to solve any homogeneous 2×2 linear system with constant coefficients. For inhomogeneous linear systems, linear systems with variable coefficients, and nonlinear systems, **dsolve** will generally not do the job. In these situations, we turn to the numerical solver **ode45**.

Suppose we wish to solve the initial value problem

$$\begin{aligned} x_1' &= F_1(t, x_1, x_2, \ldots, x_n) & x_1(t_0) &= x_{1,0} \\ x_2' &= F_2(t, x_1, x_2, \ldots, x_n) & x_2(t_0) &= x_{2,0} \\ &\vdots & &\vdots \\ x_n' &= F_n(t, x_1, x_2, \ldots, x_n) & x_n(t_0) &= x_{n,0} \end{aligned}$$

for a first order system of differential equations. This initial value problem can be written in vector form as follows:

$$\mathbf{x}'(t) = \mathbf{F}(t, \mathbf{x}), \quad \mathbf{x}(t_0) = \mathbf{x}_0,$$

where

$$\mathbf{x} = \begin{pmatrix} x_1(t) \\ x_2(t) \\ \vdots \\ x_n(t) \end{pmatrix}, \quad \mathbf{F}(t, \mathbf{x}) = \begin{pmatrix} F_1(t, x_1, \ldots, x_n) \\ F_2(t, x_1, \ldots, x_n) \\ \vdots \\ F_n(t, x_1, \ldots, x_n) \end{pmatrix}, \quad \text{and} \quad \mathbf{x}_0 = \begin{pmatrix} x_{1,0} \\ x_{2,0} \\ \vdots \\ x_{n,0} \end{pmatrix}.$$

We can use **ode45** on this system just as we use it on a single first order equation, provided the right-hand side, $\mathbf{F}$, is defined (either as an anonymous function or in an M-file) as a column vector. We illustrate with the initial value problem

$$x' = x - 2y, \quad y' = -x; \qquad x(0) = 2, \quad y(0) = -2$$

on the interval $0 \le t \le 2$. We think of x and y as coordinates of a vector $\mathbf{x}$; in MATLAB the coordinates of a vector **x** are written **x(1)** and **x(2)**.

```
>> f = @(t, x) [x(1) - 2*x(2); -x(1)];
>> [t, xa] = ode45(f, [0:0.2:2], [2 -2]);
```

The result will be an 11×1 matrix (a column vector) **t** containing the time values $0, 0.2, \ldots,$ 2, and an 11×2 matrix **xa**. The first column of **xa** contains the approximate values of $x(0)$, $x(0.2)$, ..., $x(2)$, and the second column contains the approximate values of $y(0)$, $y(0.2)$, ..., $y(2)$. These values can be displayed in a table by typing **[t xa]**. The result is shown below:

t	$x(t)$	$y(t)$
0	2.0000	-2.0000
0.2000	3.4324	-2.5349
0.4000	5.4879	-3.4143
0.6000	8.4878	-4.7927
0.8000	12.9086	-6.9036
1.0000	19.4590	-10.0974
1.2000	29.1945	-14.8984
1.4000	43.6882	-22.0907
1.6000	65.2858	-32.8448
1.8000	97.4857	-48.9081
2.0000	145.5056	-72.8881

Next we show the commands to solve numerically the system

$$x' = x + 2y, \quad y' = -x$$

and to plot a single trajectory with initial conditions $x(0) = 1$, $y(0) = 0$ on the interval $0 \leq t \leq 3$.

```
>> f = @(t, x) [x(1) + 2*x(2); -x(1)];
>> [t, xa] = ode45(f, [0 3], [1 0]);
>> plot(xa(:,1), xa(:,2))
```

Here we used **xa(:,1)** to denote the first column of **xa** (which contains the approximate values for x) and **xa(:,2)** for the second column (which contains the approximate values for y).

To plot a phase portrait for the system

$$x' = x - 2y, \quad y' = -x$$

we could use the following sequence of commands:

```
>> figure; hold on
>> f = @(t, x) [x(1) - 2*x(2); -x(1)];
>> for a = -2:2
       for b = -2:2
           [t, xa] = ode45(f, [0 3], [a b]);
           plot(xa(:,1), xa(:,2))
           [t, xa] = ode45(f, [0 -3], [a b]);
           plot(xa(:,1), xa(:,2))
       end
   end
>> axis([-20 20 -15 15])
```

These commands will produce a phase portrait similar to Figure 13.2.

Though we have used linear systems as examples in this section, numerical methods are most useful for nonlinear systems, for which solutions generally cannot be found by symbolic methods. The syntax for using **ode45** on nonlinear systems is exactly the same as in the examples above; only the definition of **f** need be changed according to the particular system. Higher order differential equations are handled by converting them into a first order system, as explained in Section 13.2, and by applying **ode45** to the resulting system. In Section 10.1 we showed how to numerically solve a second order equation in this manner.

Chapter 14

Qualitative Theory for Systems of Differential Equations

In this chapter, we extend the qualitative theory of autonomous differential equations from single equations to systems of two equations. Thus we consider a system of the form

$$\begin{cases} x' = F(x, y) \\ y' = G(x, y). \end{cases} \tag{14.1}$$

We assume that the functions F and G have continuous partial derivatives everywhere in the plane, and that the critical points—the common zeros of F and G—are isolated.

We know from the fundamental existence and uniqueness theorems that, corresponding to any choice of initial data (x_0, y_0), there is a unique pair of functions $x(t)$, $y(t)$ satisfying the system (14.1) of differential equations and the initial conditions $x(0) = x_0$, $y(0) = y_0$. Taken together, this pair of functions parameterizes a curve, or trajectory, in the *phase plane* passing through the point (x_0, y_0). Since we are dealing with autonomous systems (*i.e.*, the variable t does not appear on the right side of (14.1)), the solution curves are *independent of the starting time*. That is, if we consider initial conditions $x(t_0) = x_0$, $y(t_0) = y_0$, then we get the same solution curve with a time delay of t_0 units. Moreover, we also know that two distinct trajectories cannot intersect. Thus the plane is covered by the family of trajectories; the plot of these curves is the *phase portrait* of (14.1).

As we know, we can find explicit formula solutions $x(t)$, $y(t)$ only for simple systems. Thus we are forced to turn to qualitative and numerical methods. It is our purpose here to discuss qualitative techniques for studying the solutions of (14.1).

In most cases of physical interest, every solution curve behaves in one of the following ways:

(i) The curve consists of a single critical point.

(ii) The curve tends to a critical point as $t \to \infty$.

(iii) The curve is unbounded: the distance from $(x(t), y(t))$ to the origin becomes arbitrarily large as $t \to \infty$, or as $t \to t^*$, for some finite t^*.

(iv) The curve is periodic: the parametric functions satisfy $x(t + t_p) = x(t)$, $y(t + t_p) = y(t)$ for some fixed t_p.

(v) The curve approaches a periodic solution, *e.g.*, spiraling in on a circle.

Qualitative techniques may be able to identify the kinds of solutions that appear. In order to keep matters simple, we try to answer the following specific questions:

(a) What can we say about the critical points? Can we make a qualitative guess about their nature and stability? Can we deduce anything about the solution curves that start out close to the critical points?

(b) Can we predict anything about the long-term behavior of the solution curves? This includes those near critical points, as well as the solution curves in general.

The list of possible limiting behaviors of solution curves is based on the Poincaré-Bendixson Theorem (Theorem 9.7.3 in Boyce & DiPrima), which is valid for autonomous systems of two equations. Systems of three or more equations can have much more complicated limiting behavior. For example, solution curves can remain bounded and yet fail to approach any equilibrium or periodic state as $t \to \infty$. This phenomenon, called "chaos" by mathematicians, was anticipated by Poincaré (and perhaps even earlier by the physicist Maxwell), but most scientists did not appreciate how widespread chaos is until the arrival of computers.

There are two qualitative methods for analyzing systems of equations: one based on the idea of a *vector field*; and the other based on *linearized stability analysis*. The latter method is treated in detail in most differential equations texts; it provides information about the stability of critical points of a nonlinear system by studying associated linear systems. In this chapter, we focus on the former method: the use of vector fields in qualitative analysis. Some of the problems in Problem Set F involve linearized stability analysis; see the solution to Problem 6 in the *Sample Solutions* for an example that uses MATLAB for this analysis.

Here is the basic scenario for vector field analysis. For any solution curve $(x(t), y(t))$ and any point t, we have the differential equations

$$\begin{cases} x'(t) = F(x(t), y(t)) \\ y'(t) = G(x(t), y(t)). \end{cases}$$

If we employ vector notation

$$\mathbf{x}(t) = \begin{bmatrix} x(t) \\ y(t) \end{bmatrix}, \quad \mathbf{f}(\mathbf{x}) = \begin{bmatrix} F(\mathbf{x}) \\ G(\mathbf{x}) \end{bmatrix},$$

then the system is written as

$$\mathbf{x}' = \mathbf{f}(\mathbf{x}).$$

Moreover, the vector $\mathbf{x}'(t) = \mathbf{f}(\mathbf{x}(t))$ is just the tangent vector to the curve $\mathbf{x}(t)$. Since knowledge of the collection of tangent vectors to the solution curves gives a good idea of the curves themselves, it would be useful to have a plot of these vectors. Thus, corresponding to any vector $\mathbf{x}$, we draw the vector $\mathbf{f}(\mathbf{x})$, translated so that its foot is at the point $\mathbf{x}$. This plot is called the vector field of the system (14.1). Since we cannot do this at every point of the phase plane, we only draw the vectors at a set of regularly chosen points. Then we step back and look at the resulting vector field. We do not have the solution curves sketched, but we do have a representative collection of their tangent vectors. From the tangent vectors, we can get a good idea of the curves themselves. Indeed, we can often answer the questions above by carefully examining the vector field.

Vector fields can be drawn by hand for simple systems, but the command **quiver**, used in conjunction with several other commands, can draw vector fields of any first order system of two equations. The use of **quiver** to draw vector fields is similar to its use to draw direction fields of single first order equations (see Chapter 6). We illustrate the drawing of vector fields by examining two specific systems, namely

$$\begin{cases} x' = x(1 - x - y) \\ y' = y(0.75 - 0.5x - y); \end{cases} \tag{14.2}$$

and

$$\begin{cases} x' = x(5 - x - y) \\ y' = y(-2 + x). \end{cases} \tag{14.3}$$

System (14.2) is a competing species model, which is discussed in Boyce & DiPrima, Example 1, Section 9.4. System (14.3) is a predator-prey model with logistic growth (for the prey), which is discussed in Boyce & DiPrima, Problem 13, Section 9.5.

The following sequence of commands draws a direction field for system (14.2):

```
>> [X, Y] = meshgrid(0:1/15:2, 0:1/15:1);
>> U = X.*(1 - X - Y);
>> V = Y.*(0.75 - 0.5*X - Y);
>> quiver(X, Y, U, V)
>> axis equal tight
>> xlabel 'x'
>> ylabel 'y'
```

The result is shown in Figure 14.1. Note that we used the increment $1/15$ for x and y, resulting in a 30×15 grid. When drawing vector fields, you should generally choose the increment so that the grid has between 15 and 30 points in each direction.

In a vector field, the vector drawn at the point $\mathbf{x} = (x, y)$ indicates both the direction and length of $\mathbf{f}(\mathbf{x})$. The vector at $\mathbf{x}$ drawn by **quiver** faithfully represents the direction of $\mathbf{f}(\mathbf{x})$, but the length of each vector is rescaled so that the longest vector is about as long as the distance between neighboring points. The critical, or equilibrium, points are those points at which the vector field vanishes, *i.e.*, those points where $F(x, y) = G(x, y) = 0$. Thus the vectors are very short near the critical points. In Figure 14.1 we can approximately identify the critical points, but it is difficult to see the directions of the very short vectors near the critical points. An alternative procedure is to normalize the vectors to have the

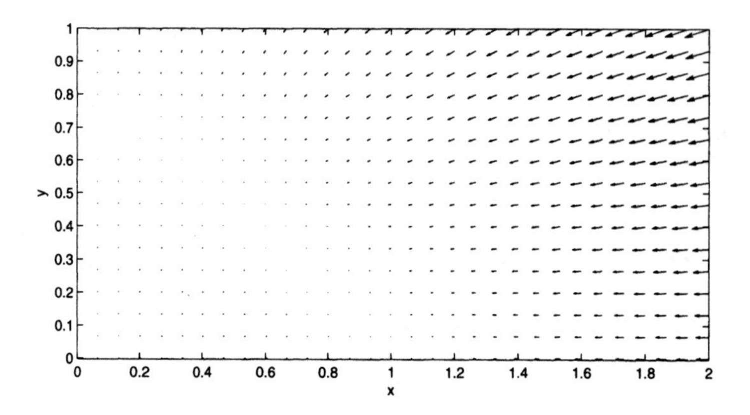

Figure 14.1: Vector Field for System (14.2)

same length, which we choose to be 0.4 times the spacing between grid points. This is done with the following sequence of commands:

```
>> [X, Y] = meshgrid(0:1/15:2, 0:1/15:1);
>> U = X.*(1 - X - Y);
>> V = Y.*(0.75 - 0.5*X - Y);
>> L = sqrt(U.^2 + V.^2);
>> quiver(X, Y, U./L, V./L, 0.4)
>> axis equal tight
>> xlabel 'x'
>> ylabel 'y'
```

Note that this sequence of commands produces a "Divide by zero" warning. This warning is caused by the fact that some of the vectors $\mathbf{f}(\mathbf{x})$ are zero; hence some of the components of **L** are zero, and some of the divisions called for in the **quiver** command are zero divided by zero. In spite of this warning, the vector field plot is correct. The result is shown in Figure 14.2.

This picture gives us a much clearer indication of the directions of the vectors. Now, however, it is harder to identify the critical points. The critical points can be found by solving the simultaneous equations $x' = 0$ and $y' = 0$. This can be done with **solve**:

```
>> syms x y
>> [x, y] = solve(x*(1 - x - y), y*(3/4 - x/2 - y)); [x y]
ans =
[   0,   0]
[   1,   0]
[   0, 3/4]
[ 1/2, 1/2]
```

Thus the critical points are $(0, 0)$, $(0, 0.75)$, $(1, 0)$, and $(0.5, 0.5)$. Knowing the critical points, we can deduce from the vector field that $(0, 0)$ is an unstable node, $(0, 0.75)$ and

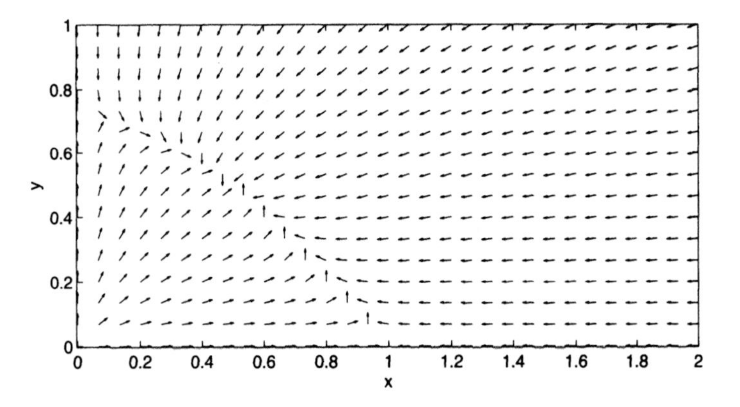

Figure 14.2: Improved Vector Field for System (14.2)

$(1, 0)$ are unstable saddle points, and $(0.5, 0.5)$ is an asymptotically stable node. Solutions starting near the saddle points (but not on the axes) tend away from them, and those starting near the point $(0.5, 0.5)$ tend toward it. Furthermore, the vector field strongly suggests that every solution curve starting in the first quadrant (but not on the axes) tends toward $(0.5, 0.5)$. Hence $(0.5, 0.5)$, which corresponds to equal populations, is apparently the limiting state that all positive solutions approach as t increases. We have thus answered both questions (a) and (b) from the beginning of this chapter on the basis of the vector field. Figure 14.3 provides a closer look at the vector field near the point $(0.5, 0.5)$.

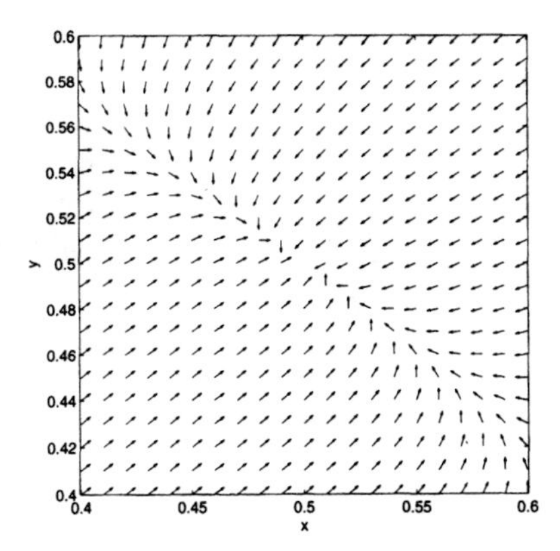

Figure 14.3: Blow-up of Vector Field for System (14.2)

Now consider the predator-prey with logistic growth system (14.3). Here $x(t)$ represents the population of prey and $y(t)$ the population of predators. The critical points of this system are $(0,0)$, $(5,0)$, and $(2,3)$. As in system (14.2), our goal is to use the vector field to answer questions (a) and (b). Here is the appropriate sequence of MATLAB commands:

```
>> [X, Y] = meshgrid(0:0.2:6, 0:0.2:4);
>> U = X.*(5 - X - Y);
>> V = Y.*(-2 + X);
>> L = sqrt(U.^2 + V.^2);
>> quiver(X, Y, U./L, V./L, 0.4)
>> axis equal tight
>> xlabel 'x'
>> ylabel 'y'
```

The output is shown in Figure 14.4.

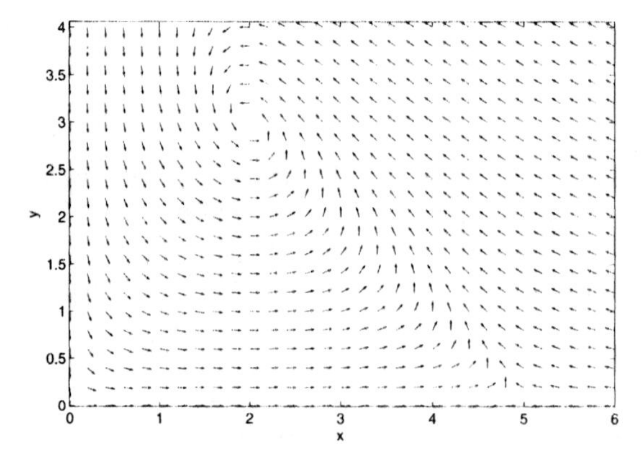

Figure 14.4: Vector Field for System (14.3)

It appears that $(0,0)$ and $(5,0)$ are unstable saddle points and that solutions spiral around $(2,3)$. The vectors near $(2,3)$ suggest that the solutions spiral into $(2,3)$, and linearized stability analysis (*cf.* the solution to Problem 6 in Problem Set F in the *Sample Solutions*) confirms that $(2,3)$ is an asymptotically stable spiral point. In particular, most solutions starting near the first two critical points tend away from them, whereas solution curves starting near the latter tend toward it. Furthermore, the vector field strongly suggests that every solution curve starting in the first quadrant (but not on the axes) tends toward the critical point $(2,3)$. This is the strongest kind of stable equilibrium—namely, no matter what the initial populations are, as long as they are both positive, the population tends toward the equilibrium position of 60% predators and 40% prey.

Remarks

- We indicated above that vector fields could be drawn by hand. It would be difficult, however, to draw as satisfactory a vector field as we have produced with MATLAB.

On the other hand, we note that careful study of the signs of F and G in system (14.1) can lead to a good understanding of the nature of the vector field.

- The essential idea in Chapter 6 for a single autonomous first order equation

$$x' = f(x)$$

is to determine the qualitative nature of the solutions purely on the basis of the sign of $f(x)$ for various x's. Actually, we have done the same thing here: namely, we have determined the qualitative nature of the solutions (trajectories) of a system from information on the signs of F and G.

- As mentioned above, according to linearized stability analysis, the nature of critical points for nonlinear systems can be deduced from the associated linear systems. If a critical point is asymptotically stable for the associated linear system (the case if the eigenvalues are negative or have negative real parts), then the point is asymptotically stable for the original nonlinear system. If it is unstable for the linear system, it is unstable for the nonlinear system. Centers are ambiguous. This analysis is valid in exactly the same way for a single equation. Suppose $\bar{x}$ is a critical point for $x' = f(x)$. Then letting $u(t) = x(t) - \bar{x}$, we see that

$$u' = f(u + \bar{x}) \approx f(\bar{x}) + f'(\bar{x})u = f'(\bar{x})u,$$

so the associated linear equation is $u' = f'(\bar{x})u$. The solutions are $u(t) = ce^{\lambda t}$, where $\lambda = f'(\bar{x})$. Therefore, $x(t) \approx \bar{x} + ce^{\lambda t}$. It is clear that $\bar{x}$ is asymptotically stable if $\lambda < 0$ and unstable if $\lambda > 0$. The same is true of the nonlinear system (with the situation for $\lambda = 0$ being ambiguous). Thus we see that this assessment of stability parallels the situation for systems.

Conclusions

You can calculate solutions of nonlinear systems and plot their trajectories with **ode45**. We described ways to do this in Chapter 13. Taken together, the information provided by a vector field, by a plot of the trajectories, and by linearized stability analysis gives a fairly complete understanding of the behavior of the solutions of a system of two first order equations. The information provided by these three approaches is similar, but each approach yields information the other approaches don't provide. For example, a vector field provides a good global indication of what the phase portrait will look like, and indicates the approximate locations of the critical points and their types; a plot of the trajectories gives a precise indication of the solution curves; and a linearized stability analysis determines the type of each critical point.

In using these approaches to study a system, it is almost always best to start with a vector field plot or with linearized stability analysis. Typically, it takes a long time to plot trajectories of nonlinear systems. But by using the information from a vector field plot or from linearized stability analysis, you can choose appropriate initial data points to use with **ode45**. A vector field plot will also help you choose an appropriate time interval (*e.g.*, positive or negative time).

Problem Set F

Systems of Differential Equations

The solution to Problem 6 appears in the *Sample Solutions*. In Chapters 13 and 14 above you can find a number of suggestions for using MATLAB to solve these problems. The parameters in those suggestions may need to be modified according to the problem at hand. In particular, you may have to change the time interval and/or the set of initial conditions. Also, be aware that MATLAB may take a long time to generate phase portraits, especially when using **ode45**. You should start with a relatively short time interval and a few initial conditions to get an idea of what the phase portrait looks like. Then, based on what you see, you should increase the time interval and/or the number of initial conditions as appropriate to improve the picture.

1. In this problem, we study three systems of equations taken from Boyce & DiPrima:

$$\mathbf{x}' = \begin{pmatrix} 2 & -1 \\ 3 & -2 \end{pmatrix} \mathbf{x} \qquad \text{(Prob. 3, Sect. 7.5)},$$

$$\mathbf{x}' = \begin{pmatrix} 1 & -1 \\ 5 & -3 \end{pmatrix} \mathbf{x} \qquad \text{(Prob. 5, Sect. 7.6)},$$

$$\mathbf{x}' = \begin{pmatrix} -3 & 5/2 \\ -5/2 & 2 \end{pmatrix} \mathbf{x} \qquad \text{(Prob. 4, Sect. 7.8)}.$$

   (a) Use MATLAB to find the eigenvalues and eigenvectors of each linear system. Use your results to write down the general solution of the system.

   (b) For each system, find the general solution using **dsolve**, and plot several trajectories. On your graphs, draw the eigenvectors (if relevant), and indicate the direction of increasing time on the trajectories. State the type and stability of the origin as a critical point. You may find that the quality of your portraits can be enhanced by modifying the t-interval or the range of values assumed by the initial data.

(c) For the first system only, compare the general solution you obtained in part (a) with the solution generated by **dsolve**. Are they the same? If not, explain how they are related (that is, match up the constants).

2. Do Problem 1 but replace the three systems listed there by the following three from Boyce & DiPrima:

$$\mathbf{x}' = \begin{pmatrix} -2 & 1 \\ 1 & -2 \end{pmatrix} \mathbf{x} \qquad \text{(Prob. 5, Sect. 7.5)},$$

$$\mathbf{x}' = \begin{pmatrix} 2 & -5/2 \\ 9/5 & -1 \end{pmatrix} \mathbf{x} \qquad \text{(Prob. 4, Sect. 7.6)},$$

$$\mathbf{x}' = \begin{pmatrix} 3 & -4 \\ 1 & -1 \end{pmatrix} \mathbf{x} \qquad \text{(Prob. 1, Sect. 7.8)}.$$

3. Use **eig** to find the eigenvalues and eigenvectors of the following systems. Use the eigenvalues and eigenvectors to write down the general solutions.

(a)

$$\mathbf{x}' = \begin{pmatrix} 3 & -1 \\ 1 & -2 \end{pmatrix} \mathbf{x}.$$

Let $\mathbf{x}(t) = (x(t), y(t))$. Determine the possible limiting behavior of $x(t)$, of $y(t)$, and of $x(t)/y(t)$ as t approaches $+\infty$.

(b)

$$\mathbf{x}' = \begin{pmatrix} 3 & 3 \\ -3 & -2 \end{pmatrix} \mathbf{x}.$$

(c)

$$\mathbf{x}' = \begin{pmatrix} -2 & -1 & 2 \\ 0 & 4 & 5 \\ 0 & -1 & 0 \end{pmatrix} \mathbf{x}.$$

Find the solution with initial condition

$$\mathbf{x}(0) = \begin{pmatrix} 7 \\ 5 \\ 5 \end{pmatrix}.$$

(d) Now solve the initial value problem in (c) with **dsolve** and compare the answer with the solution you obtained in (c).

4. MATLAB can solve some inhomogeneous linear systems. Here are three such systems (taken from Boyce & DiPrima, Sect. 7.9, Problems 2, 5, & 11):

$$\mathbf{x}' = \begin{pmatrix} 1 & \sqrt{3} \\ \sqrt{3} & -1 \end{pmatrix} \mathbf{x} + \begin{pmatrix} e^t \\ \sqrt{3}e^{-t} \end{pmatrix}, \qquad \text{(F.1)}$$

$$\mathbf{x}' = \begin{pmatrix} 4 & -2 \\ 8 & -4 \end{pmatrix} \mathbf{x} + \begin{pmatrix} t^{-3} \\ -t^{-2} \end{pmatrix}, \tag{F.2}$$

$$\mathbf{x}' = \begin{pmatrix} 2 & -5 \\ 1 & -2 \end{pmatrix} \mathbf{x} + \begin{pmatrix} 0 \\ \cos t \end{pmatrix}. \tag{F.3}$$

(a) Solve the three systems using **dsolve**.

(b) Now impose the initial data

$$\mathbf{x}(t_0) = \begin{pmatrix} 1 \\ 1 \end{pmatrix}$$

on each system, where $t_0 = 0$ in (F.1) and (F.3) and $t_0 = 1$ in (F.2). For each of (F.1)–(F.3), draw the solution curve using a time interval $-1 \leq t \leq 1$ in (F.1) and (F.3), and the interval $\frac{1}{2} \leq t \leq \frac{3}{2}$ in the other case. Based on the differential equation, indicate the direction of motion through the initial value. Then expand the time interval and, purely on the graphical evidence, venture a guess as to the behavior of the solution curve as $t \to \infty$ and as $t \to -\infty$ ($t \to 0$ in (F.2)).

5. Here we reconsider the pendulum models examined in Problems 3–6 of Problem Set D. (See also the discussion in Sections 9.2 and 9.3 of Boyce & DiPrima.)

(a) Consider first the undamped pendulum

$$\theta'' + \sin\theta = 0, \quad \theta(0) = 0, \quad \theta'(0) = b.$$

Let $x = \theta$ and $y = \theta'$; then x and y satisfy the system

$$\begin{cases} x' = y \\ y' = -\sin x, \end{cases} \qquad \begin{cases} x(0) = 0 \\ y(0) = b. \end{cases}$$

Solve this system numerically and plot, on a single graph, the resulting trajectories for the initial velocities $b = 0.5, 1, 1.5, 2, 2.5$. Use a positive time range (*e.g.*, $0 \leq t \leq 15$), and use **axis equal** to improve your pictures.

(b) Based on your pictures in part (a), describe physically what the pendulum seems to be doing in the three cases $\theta'(0) = 1.5, 2$, and 2.5.

(c) The energy of the pendulum is defined as the sum of kinetic energy $(\theta')^2/2$ and potential energy $1 - \cos\theta$, so

$$E = \frac{1}{2}(\theta')^2 + 1 - \cos\theta = \frac{1}{2}y^2 + 1 - \cos x.$$

Show, by taking the derivative dE/dt, that E is constant when $\theta(t)$ is a solution to the equation. What basic physical principle does this represent?

(d) Use **contour** to plot the level curves for the energy. Choose the x and y ranges according to where most of the solutions from part (a) lie, and use the optional fourth argument to **contour** to increase the number of level curves plotted. Explain how your picture is related to the trajectory plot from part (a).

(e) If the pendulum reaches the upright position (*i.e.*, $\theta = \pi$), what must be true of the energy? Now explain why there is a critical value E_0 of the energy, below which the pendulum swings back and forth without reaching the upright position, and above which it swings overhead and continues to revolve in the same direction. What is the critical value E_0? What is the value of b corresponding to E_0? What do you expect the pendulum to do when b has this critical value? What did the plot in part (a) show? Explain.

(f) Now consider the damped pendulum

$$\theta'' + 0.5\theta' + \sin\theta = 0, \quad \theta(0) = 0, \quad \theta'(0) = b.$$

Numerically solve the corresponding first order system (which you must determine) and plot the resulting trajectories for the initial velocities $b = 0.5$, 1, 1.5, ..., 6. You will notice that each of the trajectories in your graph tends toward either the critical point $(0, 0)$ or the critical point $(2\pi, 0)$. Explain what this means physically. There is a value b_0 for which the trajectory tends toward the critical point $(\pi, 0)$. Estimate, up to two-decimal accuracy, the value of b_0. What would this correspond to physically?

(g) Recall the energy defined in part (c). Compute E' and determine which of the following are possible: (1) E is increasing, (2) E is decreasing, (3) E is constant. Explain how the possibilities are reflected in the solutions to part (f).

6. Consider the *competing species* model (Boyce & DiPrima, Prob. 2, Sect. 9.4)

$$\begin{cases} \frac{dx}{dt} = x(1.5 - x - 0.5y) \\ \frac{dy}{dt} = y(2 - 1.5x - 0.5y) \end{cases}$$

for $x, y \geq 0$.

(a) Find all critical points of the system. At each critical point, calculate the corresponding linear system and find the eigenvalues of the coefficient matrix; then identify the type and stability of the critical point.

(b) Plot the vector field on a region small enough to distinguish the critical points but large enough to judge the possible solution behaviors away from the critical points.

(c) Use several initial data points (x_0, y_0) in the first quadrant to draw a phase portrait for the system. Identify the direction of increasing t on the trajectories you obtain. Use the information from parts (a) and (b) to choose a representative sample of initial conditions. Then combine the vector field and phase portrait on a single graph.

(d) Suppose the initial state of the population is given by

$$x(0) = 2.5,\ y(0) = 2.$$

Find the state of the population at $t = 1, 2, 3, 4, 5, \ldots, 20$.

(e) Explain why, practically speaking, there is no "peaceful coexistence"; *i.e.*, with the exception of an atypical set of starting populations (the *separatrix* curves), one or the other population must die out. For which nonzero initial populations is there no change? Add to your final plot from part (c) the separatrices, *i.e.*, the solution curves that approach the unstable equilibrium point where both populations are positive. These curves form the boundary between the solution curves that approach each of the two stable points where only one population survives. *Hint*: Since a separatrix is a solution curve that *approaches* a saddle point, you can obtain it by plotting trajectories starting very close to the saddle point and going *backwards* in time.

(f) The vertical line $x = 1.5$ cuts the separatrix. By using the event detection feature of **ode45** and the hint in part (e), find a numerical approximation to the number $\bar{y}$ such that $(1.5, \bar{y})$ is on the separatrix.

7. Consider the *competing species* model (Boyce & DiPrima, Prob. 4, Sect. 9.4)

$$\begin{cases} \frac{dx}{dt} = x(1.5 - 0.5x - y) \\ \frac{dy}{dt} = y(0.75 - 0.125x - y). \end{cases}$$

for $x, y \geq 0$.

(a) Find all critical points of the system. At each critical point, calculate the corresponding linear system and find the eigenvalues of the coefficient matrix; then identify the type and stability of the critical point.

(b) Plot the vector field on a region small enough to distinguish the critical points but large enough to judge the possible solution behaviors away from the critical points.

(c) Use several initial data points (x_0, y_0) in the first quadrant to draw a phase portrait for the system. Identify the direction of increasing t on the trajectories you obtain. Use the information from parts (a) and (b) to choose a representative sample of initial conditions. Then combine the vector field and phase portrait on a single graph.

(d) Suppose the initial state of the population is given by

$$x(0) = 0.1,\ y(0) = 0.1.$$

Find the state of the population at $t = 1, 2, 3, 4, 5, \ldots, 20$.

(e) Explain why, practically speaking, "peaceful coexistence" is the only outcome; *i.e.*, with the exception of the situation in which one or both species starts out without any population, the population distributions always tend toward a certain equilibrium point. Sketch on your final plot from part (c) the separatrices that connect the stable equilibrium point to the two unstable points at which one population is zero; these separatrices divide the solution curves that tend toward the origin as $t \to -\infty$ from those that are unbounded as $t \to -\infty$. *Hint*: Since the separatrices are trajectories coming *out* of the saddle points, one can plot them by solving the equation for initial conditions very close to the saddle points.

(f) The vertical line $x = 2.5$ cuts a separatrix. By using the event detection feature of **ode45** and the hint in part (e), find a numerical approximation to the number $\bar{y}$ such that $(2.5, \bar{y})$ is on the separatrix.

8. Consider the *predator-prey* model

$$\begin{cases} \frac{dx}{dt} = x(4 - 3y) \\ \frac{dy}{dt} = y(x - 2) \end{cases}$$

in which $x \geq 0$ represents the population of the prey and $y \geq 0$ represents the population of the predators.

(a) Find all critical points of the system. At each critical point, calculate the corresponding linear system and find the eigenvalues of the coefficient matrix; then identify the type and stability of the critical point.

(b) Plot the vector field on a region small enough to distinguish the critical points but large enough to judge the possible solution behaviors away from the critical points.

(c) Use several initial data points (x_0, y_0) in the first quadrant to draw a phase portrait for the system. Identify the direction of increasing t on the trajectories you obtain. Use the information from parts (a) and (b) to choose a representative sample of initial conditions. Then combine the vector field and phase portrait on a single graph.

(d) Explain from your phase portrait how the populations vary over time for initial data close to the unique critical point inside the first quadrant. What happens for initial data far from this critical point?

(e) Suppose the initial state of the population is given by

$$x(0) = 1, \; y(0) = 1.$$

Find the state of the population at $t = 1, 2, 3, 4, 5$.

(f) Estimate to two decimal places the period of the solution curve that starts at $(1, 1)$.

9. Consider a *predator-prey* model where the behavior of the prey is governed by a logistic equation (in the absence of the predator). Such a model is typified by the following system (Boyce & DiPrima, Prob. 3, Sect. 9.5):

$$\begin{cases} \frac{dx}{dt} = x(1 - 0.5x - 0.5y) \\ \frac{dy}{dt} = y(-0.25 + 0.5x), \end{cases}$$

where $x \geq 0$ represents the population of the prey and $y \geq 0$ represents the population of the predators.

(a) Find all critical points of the system. At each critical point, calculate the corresponding linear system and find the eigenvalues of the coefficient matrix; then identify the type and stability of the critical point.

(b) Plot the vector field on a region small enough to distinguish the critical points but large enough to judge the possible solution behaviors away from the critical points.

(c) Use several initial data points (x_0, y_0) in the first quadrant to draw a phase portrait for the system. Identify the direction of increasing t on the trajectories you obtain. Use the information from parts (a) and (b) to choose a representative sample of initial conditions. Then combine the vector field and phase portrait on a single graph.

(d) Explain from your phase portrait how the populations vary over time for initial data close to the unique critical point inside the first quadrant. What happens for initial data far from this critical point?

(e) Suppose the initial state of the population is given by

$$x(0) = 1,\ y(0) = 1.$$

Find the state of the population at $t = 1, 2, 3, 4, 5$.

(f) Estimate how long it takes for both populations to be within 0.01 of their equilibrium values if we start with initial data $(1, 1)$.

10. Consider a modified predator-prey system where the behavior of the prey is governed by a logistic/threshold equation (in the absence of the predator). Such a model is typified by the following system (Boyce & DiPrima, Prob. 5, Sect. 9.5):

$$\begin{cases} \frac{dx}{dt} = x(-1 + 2.5x - 0.3y - x^2) \\ \frac{dy}{dt} = y(-1.5 + x) \end{cases}$$

where $x \geq 0$ represents the population of the prey, and $y \geq 0$ represents the population of the predators.

(a) Find all critical points of the system. At each critical point, calculate the corresponding linear system and find the eigenvalues and eigenvectors of the coefficient matrix; then identify the critical points as to type and stability.

(b) Plot the vector field on a region small enough to distinguish the critical points but large enough to judge the possible solution behaviors away from the critical points.

(c) Use several initial data points (x_0, y_0) in the first quadrant to draw a phase portrait for the system. Identify the direction of increasing t on the trajectories you obtain. Use the information from parts (a) and (b) to choose a representative sample of initial conditions. Then combine the vector field and phase portrait on a single graph.

(d) In parts (a), (b), and (c), you obtained information about the solutions of the system using three different approaches. Combine all of this information by combining the plots from parts (b) and (c), and then marking the critical points and adding any separatrices. (See the hints in Problems 6 and 7 about how to plot separatrices.) What information does the approach in part (a) provide that the other approaches do not? Answer the same question for parts (b) and (c).

(e) Interpret your conclusions in terms of the populations of the two species.

11. Some of the nonlinear systems of differential equations we have studied have more than one asymptotically stable equilibrium point. In such cases, it can be difficult to predict which (if any) equilibrium point the solution with a given initial condition will approach as time increases. Generally, the solution curves that approach one stable equilibrium will be separated from the solutions that approach another stable equilibrium by a separatrix curve. The separatrix is itself a solution curve that does not approach either stable equilibrium—often it approaches a saddle point instead. (See, for example, Figure 9.4.4 in Section 9.4 of Boyce & DiPrima.)

    Having located a relevant saddle point, one can approximate the separatrix by choosing an initial condition very close to the saddle point and solving the differential equation *backwards* in time. The reason for going "back in time" is to find (approximately) a solution curve that approaches very close to the saddle point as time increases. We will apply this idea in a population dynamics model to get a fairly precise picture of which initial conditions go to which equilibria.

    (a) Consider the system

$$\begin{cases} \frac{dx}{dt} = x(-1 + 2.5x - 0.3y - x^2) \\ \frac{dy}{dt} = y(-1.5 + x) \end{cases}$$

studied Problem 10. There are two asymptotically stable equilibria, $(0, 0)$ and $(3/2, 5/3)$, and two saddle points at $(0.5, 0)$ and $(2, 0)$. Plot a family of solution curves in the first quadrant (positive x and y) on the same graph. Use a time interval from $t = 0$ to a positive time large enough that you can clearly see where the solutions curves are headed, but not so large that your plot takes forever to compute. Choose a set of initial values robust enough to give a reasonably "thick" set of solution curves. It also may be useful to adjust the range of the plot until you get an approximately square graph. Observe where the solution curves seem to be heading and where the separatrix appears to be.

(b) Draw an approximate separatrix by plotting a solution curve with initial values for x and y very close to the saddle point (1/2,0), using a *negative* range of values for t. This should give you a good picture of the separatrix. Superimpose the separatrix on the portrait of solution curves you drew in part (a). Make sure the separatrix really does separate the different asymptotic properties of the solution curves found in part (a). (To help in distinguishing the separatrix, use a different line style, such as dashed or dotted, in making your separatrix plot. To learn the available line styles, see the help text for **plot**.) Discuss how the possible limiting values of x and y of a solution curve depend on its initial values.

12. In Section 5.3 we investigated the sensitivity to initial values of solutions of a single differential equation. In this problem, we investigate the same issue for a system of equations. The system below is said to be *chaotic* because the solutions are sensitive to initial values, but in contrast to Example 5.3 and Problems 15 and 16 of Problem Set C, the solutions remain bounded.

    Since we do not know the exact solutions for the system we study, in order to judge the time it takes for a small perturbation of the system to become large, we will compare pairs of numerical solutions whose initial conditions are close to each other. We consider perturbations of decreasing size from 10^{-1} to 10^{-4}. (Though **ode45** with the default error tolerance can make errors larger than 10^{-4}, it will make approximately the same error for two nearby solutions, so it is still a suitable tool for estimating how long it takes two nearby solutions to grow apart from each other. On the other hand, it would be reasonable to increase the accuracy of **ode45** as described in Section 7.3.)

    (a) Consider the Lorenz system

$$\begin{cases} x' = 10(y - x) \\ y' = 28x - y - xz \\ z' = -(8/3)z + xy \end{cases}$$

(which is studied, for example, in Section 9.8 of Boyce & DiPrima). We investigate the idea that small changes in the initial conditions can lead to large changes in the solution after a relatively short period of time. For $\delta = 0.1$, plot on the same graph the first coordinate $x(t)$ of the solutions corresponding to the initial conditions $(3, 4, 5)$ and $(3, 4, 5 + \delta)$ from $t = 0$ to $t = 20$. Observe the time at which the solutions start to differ substantially. Repeat for $\delta = 0.01, 0.001, 0.0001$. (The criterion for a "substantial" difference between the two solutions is up to you; just try to be consistent from one observation to the next.)

(b) Make a table or graph of the number of decimal places in which the initial condition was perturbed (that is, 1, 2, 3, and 4 for the four given values of δ) versus the amount of time the solutions stayed close to each other. This graph shows roughly how long we should trust a numerical solution to be close to the actual solution for a given number of digits of accuracy in each step of our numerical method. Describe how the amount of time we can trust a numerical solution seems to depend on the number of digits of accuracy per time step, judging from your data. If a numerical method has 14 digits of accuracy, about how long do you think the numerical solution can be trusted? Roughly how many digits of accuracy would be needed to trust the solution up to $t = 1000$?

13. Read the introduction to Problem 12.

(a) Do part (a) of Problem 12, but using the Rössler system

$$\begin{cases} x' = -y - z \\ y' = x + 0.36y \\ z' = 0.4x - 4.5z + xz. \end{cases}$$

This time plot $y(t)$ from $t = 0$ to $t = 100$ and use $(2, 2, 2)$ as the initial condition in place of $(3, 4, 5)$.

(b) Do part (b) of Problem 12 for the Rössler system, this time initially assuming a numerical method accurate to 10 digits.

14. A simple model of a nonlinear network, of a sort that comes up for example in modeling portions of the nervous system, consists of a number n of identical units connected together, with the state x_j of the j-th unit satisfying the differential equation

$$x_j' = f(x_j - x_{j-1}), \tag{F.4}$$

meaning that each unit changes by a fixed function f of the difference between the state of the unit and the state of its predecessor. (See Figure F.1 for a schematic diagram.) For the sake of this problem, take $x_0 = x_n$. (That means the network

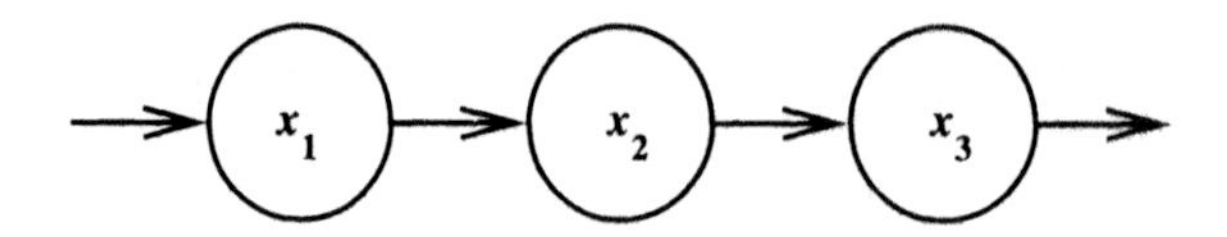

Figure F.1: A Multi-Unit Network

closes up, with the last unit providing input to the first in (F.4) with $j = 1$.) Suppose $n = 5$.

(a) The simplest interesting model of this sort is when f is linear. Since we want x_j to increase when x_{j-1} is bigger than it, that suggests taking $f(x_j - x_{j-1}) = -\lambda(x_j - x_{j-1})$ with $\lambda > 0$. In other words,

$$\mathbf{x}' = \lambda A\mathbf{x}, \qquad A = \begin{pmatrix} -1 & 0 & 0 & 0 & 1 \\ 1 & -1 & 0 & 0 & 0 \\ 0 & 1 & -1 & 0 & 0 \\ 0 & 0 & 1 & -1 & 0 \\ 0 & 0 & 0 & 1 & -1 \end{pmatrix}. \tag{F.5}$$

Have MATLAB verify that $(A + I)^5 = I$. That says that the eigenvalues of A are of the form $-1 + e^{2\pi ij/5}$, or that the eigenvalues of λA are of the form $\lambda(-1 + e^{2\pi ij/5})$, for some integers j. What does this say about the solutions to (F.5) as $t \to \infty$?

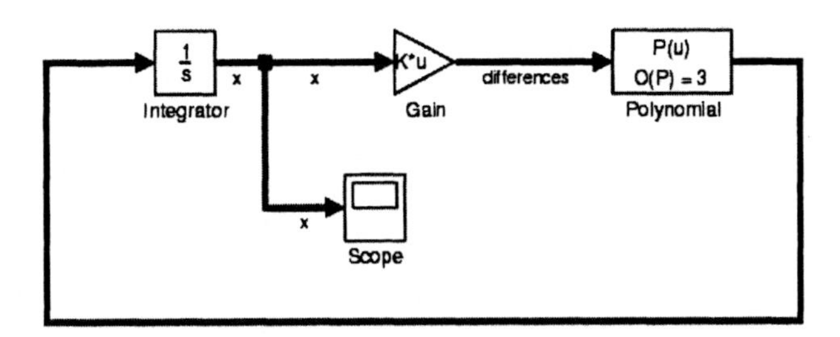

Figure F.2: Simulink Model for the Multi-Unit Network

(b) Construct a Simulink model as shown in Figure F.2. If you set the **Block Parameters** of the Gain block to "matrix multiplication" $K * x$ and set the gain K to the matrix A above, you get a model of the network where f can be an arbitrary polynomial. First set the coefficients of the polynomial in the Polynomial block to `[1, 0]`. That means 1 is the coefficient of x and 0 is the constant term, *i.e.*, you have the model of (a) with $\lambda = 1$. Try varying the initial conditions in the Integrator block, and show that you recover the results of part (a).

(c) Try setting the coefficients of the polynomial in the Polynomial block to `[-b, 1, 1, 0]`, where **`b`** varies between 0 and 1. (This corresponds to a polynomial $-bx^3+x^2+x$.) You can take the initial condition for x to be **`rand(5,1)`**, meaning a 5-dimensional column vector of random numbers between 0 and 1. Experiment with different value of b and discuss how the results compare with the linear model in (a).

15. This problem is a follow-up to Problem 14 above. One possible modification to the model in that problem is to assume that it takes a certain amount of time T

for the difference between the states of a unit and its predecessor to have an effect. (For example, if the model represents a network of neurons, it might take time for signals to propagate across the synapse between them.) For simplicity, let's consider the linear model in (F.5). The matrix A remains unchanged, but the equation now becomes a differential/difference equation

$$\mathbf{x}'(t) = \lambda A \mathbf{x}(t - T). \tag{F.6}$$

We can still find solutions to this equation of the form $\mathbf{x}(t) = e^{\sigma t}\mathbf{v}$ provided that $\mathbf{v}$ is an eigenvector of A. For if $A\mathbf{v} = \mu\mathbf{v}$, then substituting in (F.6) gives a *scalar* equation relating λ, σ, μ, and T.

(a) Find this equation. (Note that σ, μ, and $\mathbf{v}$ are allowed to be complex; this corresponds to allowing solutions involving sine and cosine functions after we take real parts.)

(b) A *standing wave* solution to (F.6) is a purely oscillatory solution, *i.e.*, one for which σ is purely imaginary. Show that such solutions exist for certain special values of the parameters.

(c) Modify the Simulink model from Problem 14 by inserting a Unit Delay block. (This amounts to choosing units of time so that $T = 1$.) See what happens as you increase λ (*i.e.*, you take the coefficients of the polynomial in the Polynomial block to be of the form **`[lambda, 0]`** with varying **`lambda`**). Things get interesting for λ around 0.3.

16. The subject of *chemical kinetics* studies the rate at which chemical reactions take place. For a reaction that takes place in one step, say

$$A + B \to AB,$$

the rate at which the reaction proceeds to the right (at least at low concentrations of the reactants) is proportional to the product of the concentrations of A and B, since formation of compound AB depends on the collision of a molecule of A with a molecule of B. (Strictly speaking, we are talking about a situation where all the substances involved are gases, and it makes sense to talk about molecular collisions. The case where A and B are dissolved in water is slightly more complicated.)

But for a reaction like

$$2A + B \to A_2B,$$

there are two possibilities. It might be that the reaction proceeds in one step, in which case the rate would be proportional to the product of the *square* of the concentration of A with the concentration of B. Such a reaction is called *second order* in A, *first order* in B. Or it might be that the reaction really proceeds in two steps,

$$A + B \to AB, \qquad A + AB \to A_2B,$$

in which case the square of the concentration of A never comes in (since there is no need for two molecules of A to collide). In this problem, we analyze the second possibility. Let x, y, z, and w be the concentrations of A, B, AB, and A_2B, respectively (in units proportional to the total number of molecules). Conservation of matter says that

$$x + z + 2w \quad \text{(total amount of } A\text{)}, \qquad y + z + w \quad \text{(total amount of } B\text{)}$$

remain constant. For definitiveness, we will assume that (again, in appropriate units)

$$x + z + 2w = 2, \qquad y + z + w = 1. \tag{F.7}$$

(a) Use **solve** to solve the equations (F.7) for w and z. Your result should be $w = 1 - x + y$, $z = x - 2y$.

(b) The two-step chemical reaction is modeled by the differential equations

$$w' = \beta x z$$
(rate of creation of A_2B is proportional to xz)
$$z' = \alpha x y - \beta x z$$
(AB is being created from A and B, destroyed by creation of A_2B)

for positive constants α and β. Use part (a) and **solve** to rewrite the differential equations in terms of x, y, x' and y'. This should yield

$$\begin{cases} x' = (2\beta - \alpha)xy - \beta x^2, \\ y' = -\alpha x y, \end{cases} \tag{F.8}$$

an autonomous system that falls into the framework of Chapter 14.

(c) Show that there is only one critical point of (F.8) in the physically realistic domain where x, y, z, and w are all ≥ 0. What is it? Explain why that critical point does not fall under the rubric of *almost linear system* theory, as developed for example in Boyce & DiPrima, Section 9.3.

(d) Nevertheless, we can uncover a great deal of qualitative information about the solution curves. Plot the vector field in the case $\alpha = 1$, $\beta = 10$. (This is the case where the intermediate product AB is relatively unstable and converts rapidly to A_2B.) What does the field look like in the physically relevant region? Superimpose several trajectories of the system on your plot, starting along the line segment joining $x = 1$, $y = 0$ and $x = 2$, $y = 1$. How do the solution curves approach the critical point?

(e) Plot the vector field in the case $\alpha = 10$, $\beta = 1$. (This is the case where the intermediate product AB is relatively stable.) Again superimpose several trajectories of the system on your plot, starting along the line segment joining $x = 1$, $y = 0$ and $x = 2$, $y = 1$. How do the results differ from the previous case?

(f) Finally, suppose the initial state of the system is $x = 2$, $y = 1$ (corresponding to $w = z = 0$, *i.e.*, no AB or A_2B). Plot the concentration w of the final product A_2B as a function of time t, in the two cases studied above ($\alpha = 1$ and $\beta = 10$, then $\alpha = 10$ and $\beta = 1$), on the same set of axes. Comment on the difference. How do the results compare with what you would get in the case where the reaction

$$2A + B \rightarrow A_2B$$

proceeds in a single step?

17. This problem continues the discussion of chemical kinetics in Problem 16. Again consider a gaseous reaction, but this time assume that it is reversible, so that we have a reaction of the form

$$A + B \leftrightharpoons AB.$$

Let x, y, and z be the concentrations of A, B, and AB, respectively. Just as in Problem 16, $x + z$ and $y + z$ remain constant in time, because of conservation of matter. Let's assume that (in suitable units) $x + z = 1$, but that $y + z$ can be bigger than 1. Since AB is being both created (at a rate proportional to the product of the concentrations of A and B) and destroyed (at a rate proportional to the concentration of AB), the differential equation becomes:

$$z' = -x' = -y' = \alpha xy - \gamma z = \alpha xy - \gamma(1 - x),$$

or

$$\begin{cases} x' = -\alpha xy + \gamma(1 - x), \\ y' = -\alpha xy + \gamma(1 - x). \end{cases} \tag{F.9}$$

Study this system in the domain $0 \leq x \leq 1, y \geq 0$.

(a) Show that (F.9) has a curve of critical points given by the equation

$$y = \frac{\gamma}{\alpha}\left(\frac{1}{x} - 1\right).$$

(b) Plot the vector field in the case $\alpha = 1$, $\gamma = 10$. (This is the case where the product AB is relatively unstable and converts back rapidly to A and B.) What does the field look like in the physically relevant region? Superimpose several trajectories of the system on your plot, starting along the line segment joining $x = 0$, $y = 0$ and $x = 0$, $y = 3$. (*Note*: $x = 0$ means no A is present initially, but then $z = 1$, so AB is present, and because of the reversible reaction, A is eventually created.)

(c) Plot the vector field in the case $\alpha = 10$, $\gamma = 1$. (This is the case where the product AB is relatively stable.) Again superimpose several trajectories of the system on your plot, starting along the line segment joining $x = 0$, $y = 0$ and $x = 0$, $y = 3$. How do the results differ from the previous case?

(d) Finally, suppose the initial state of the system is $x = 0$, $y = 2$. Plot the concentration z of the product AB as a function of time t, in the two cases studied above ($\alpha = 1$ and $\gamma = 10$, then $\alpha = 10$ and $\gamma = 1$), on the same set of axes. What do you observe?

18. A magnetic servomotor is depicted in Figure F.3. The input voltage $v(t)$ results in a current $i(t)$, which produces a torque on the rotor. The rotor, with angle $\theta(t)$, acts like a generator, and a circuit-torque analysis leads to the pair of differential equations (*cf.* E. B. Magrab, S. Azarm, B. Balachandran, J. Duncan, K. Herold, G. Walsh, *An Engineer's Guide to MATLAB*, Prentice Hall, 2001)

$$\begin{cases} Li'(t) + k_b\theta'(t) + Ri(t) = v(t) \\ J\theta''(t) + b\theta'(t) - k_\tau i(t) = 0, \end{cases} \tag{F.10}$$

where the constants are: R (motor resistance), L (inductance), J (inertia), b (motor friction), k_b (back emf generator constant), and k_τ (conversion factor from current to torque).

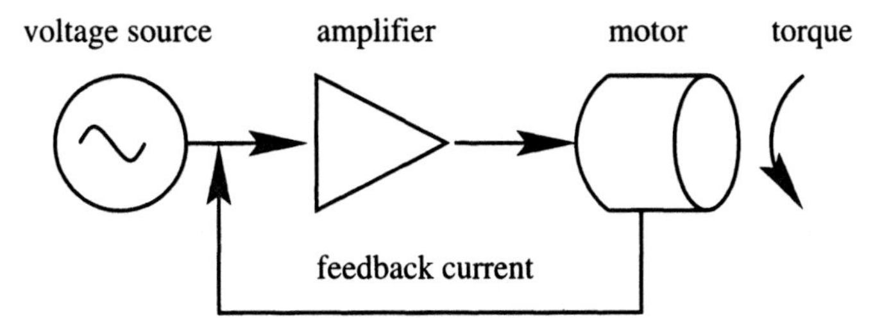

Figure F.3: Schematic Picture of a Magnetic Servomotor

(a) Convert the pair of equations (F.10) to a system of three first order linear equations.

(b) Assign values to the constants as follows (we ignore units here): $L = 0.005, R = 5.0, k_b = 0.125, k_\tau = 15, J = 0.03, b = 0.01$. Now solve the system using **dsolve**. For initial data you may take $i(0) = \theta(0) = \theta'(0) = 0$ and set the impressed voltage to be $v(t) \equiv 1$. What is the motor doing? (To answer, graph the rotor angle $\theta(t)$.) Is that realistic?

(c) Now assume there is a feedback mechanism. This means that

$$v(t) = \kappa(r(t) - \theta(t)),$$

where κ and $r(t)$ are to be specified. Take $r(t) \equiv 1$, the same vanishing initial data and $\kappa = 1$; then solve the resulting system. Now what is the motor doing? Is this realistic?

(d) Increase the value of κ to 50, then 150, and finally 500. Describe the different behaviors that you see and what they imply for the motor.

(e) A somewhat more realistic, albeit nonlinear, model is obtained if the feedback function is

$$v(t) = \tanh(\kappa(r(t) - \theta(t))).$$

Repeat part (d) with $\kappa = 1$ and 150. This time you will need to use **ode45** instead of **dsolve**. What happens in the second case is quite subtle. Can you identify the long-term behavior?

Glossary

This glossary is divided into the following sections:

1. MATLAB Operators: the special symbols used by MATLAB,
2. MATLAB Commands: commands that manipulate data or expressions, or that initiate a process,
3. Built-in Functions: basic mathematical functions that can be evaluated or plotted,
4. Graphics Commands: commands using in creating or modifying figures,
5. Built-in Constants,
6. MATLAB Programming: commands used for programming and structuring M-files,
7. Simulink Blocks: the Simulink blocks most useful for the Simulink problems in this book.

The distinction among these various categories, especially among commands and programming constructs, is somewhat artificial.

Each item is followed by a brief description of its effect and then (in most cases) one or more examples. To get more information on a command, you can use the **help** command or the Help Browser. This glossary is not a comprehensive list of MATLAB commands, but it includes the commands most useful for studying differential equations.

MATLAB Operators

@ Marker for "function handles." Also used to create anonymous functions.

```
f = @(x, y) x.^2.*y + y.^2
quadl(@atan, 0, 1)
```

**** Left matrix division. `X = A\B` is the solution of the equation `A*X = B`. Type **help slash** for more information.

```
A = [1 0; 2 1]; B = [3; 5]; A\B
```

/ Ordinary scalar division, or right matrix division. Type **help slash** for more information.

`./` Element-by-element division of arrays. See the online help for **rdivide**.

`*` Scalar or matrix multiplication. See the online help for **mtimes**.

`.*` Element-by-element multiplication of arrays. See the online help for **times**.

`^` Scalar or matrix powers. See the online help for **mpower**.

`.^` Element-by-element powers. See the online help for **power**.

`:` Range operator, used for defining vectors and matrices. Type **help colon** for more information.

`'` Complex conjugate transpose of a matrix. See **ctranspose**. Also used to begin and end strings.

`.'` Transpose of a matrix. See **transpose**.

`;` Suppresses output of a MATLAB command.

```
X = 0:0.1:30;
```

`...` Line continuation operator. Cannot be used inside quoted strings. Type **help punct** for more information.

```
1 + 3 + 5 + 7 + 9 + 11 ...
   + 13 + 15 + 17
['This is a way to create very long strings ', ...
   'that span more than one line.  Note the square ', ...
   'brackets']
```

`.` Not a true MATLAB command. Used in conjunction with arithmetic operators to force element-by-element operations on arrays.

```
a = [1 2 3]; b = [x y z]; a.*b
```

MATLAB Commands

addpath Adds the specified directory to MATLAB's file search path.

```
addpath('C:\my_mfiles')
addpath C:\my_mfiles
```

ans A variable holding the value of the most recent unassigned output.

bvp4c Boundary value problem solver. Usually requires use of **bvpinit** to set the "initial guess" argument. The example below solves and plots the solution of $y'' + y = 0$ with boundary values $y(0) = 0$, $y(\pi/2) = 1$.

```
solinit = bvpinit((0:0.1:1)*pi/2, [1, 0]);
sol = bvp4c(@(x, y) [y(2); -y(1)], ...
   @(ya, yb) [ya(1); yb(1) - 1], solinit);
plot(sol.x, sol.y(1,:))
```

cd Makes the specified directory the current (working) directory.

```
cd C:\mydocs\mfiles
```

char Converts a symbolic expression to a string. Useful for defining inline functions or for defining input to **dsolve**.

```
syms x y
f = inline(char(sin(x)*sin(y)), 'x', 'y')
```

clear Clears values and definitions for variables and functions. If you specify one or more variables, then only those variables are cleared.

```
clear
clear f, g
```

collect Collects coefficients of powers of the specified symbolic variable in a given symbolic expression.

```
syms x y
collect(x^2 - 2*x^2 + 3*x + x*y, x)
```

conj Gives the complex conjugate of a complex number.

```
conj(2 + 3*i)
```

ctranspose Conjugate transpose of a matrix. Usually invoked with the **'** operator. Equivalent to **transpose** for real matrices.

```
A = [1 3 i]; A'
```

D Not a true MATLAB command. Used in the **dsolve** command to denote differentiation. See **diff**.

```
dsolve('t*Dy + y = sin(t)')
```

deal Distributes its arguments to different variables.

```
[r, t] = deal(sqrt(x^2 + y^2), atan2(y, x))
```

delete Deletes a file from the disk.

```
delete filename
```

det The determinant of a matrix.

```
det([1 3; 4 5])
```

deval Evaluates a numerical solution to an ODE at specified points.

```
sol = ode45(@(t, y) t.*y, [0, 5], 1)
deval(sol, 1)
```

diff Symbolic differentiation operator (and difference operator).

```
syms x; diff(x^3)
syms x y; diff(x*y^2, y)
```

dir Lists the files in a directory.

disp Displays a string or the value of a variable.

```
disp('The answer is:'), disp(a)
```

doc Displays details about a specific command in the Help Browser.

```
doc print
```

double Gives a double precision value for either a numeric or symbolic quantity. Applied to a string, **double** returns a vector of ASCII codes for the characters in the string.

```
z = sym('pi'); double(z)
```

dsolve Symbolic ODE solver. By default, the independent variable is **t**, but a different variable can be specified as the last argument. The input must be a string, not a symbolic expression.

```
dsolve('D2y - t*y = 0')
dsolve('Dy + y^2 = 0', 'y(0) = 1', 'x')
[x, y] = dsolve('Dx = 2*x + y', 'Dy = -x')
```

echo Turns on or off the echoing of commands inside script M-files.

edit Opens an M-file in the Editor/Debugger.

```
edit mymfile
```

eig Computes eigenvalues and eigenvectors of a square matrix.

```
eig([2, 3; 4, 5])
[e, v] = eig([1, 0, 0; 1, 1, 1; 1, 2, 4])
```

end Last entry of a vector. (Also a programming construct.)

```
v(end)
v(3:end)
```

eval Evaluates a string or symbolic expression. Useful in M-files or in defining anonymous functions.

```
sol = dsolve('Dy = t^2 + y', 'y(0) = 2', 't');
fsol = @(t) eval(vectorize(sol))
```

eye The identity matrix of the specified size.

```
eye(5)
```

factor Factors a polynomial or integer.

```
syms x y; factor(x^4 - y^4)
```

feval Evaluates a function specified by a string. Useful in function M-files.

```
feval('exp', 1)
```

format Specifies the output format for numerical variables.

```
format short
format long
```

fzero Tries to find a zero of the specified function or expression (given as a string) near a given starting point or on a specified interval. If the first argument is a string, the independent variable must be **x**.

```
fzero(@cos, [-pi 0])
fzero('cos(x) - x', 1)
```

help Asks for documentation for a MATLAB command. See also **lookfor**.

```
help factor
```

ilaplace Inverse Laplace Transform.

```
syms s t; ilaplace(1/s, s, t)
```

inline Constructs a MATLAB inline function from a string expression. Quite useful in MATLAB 6, but discouraged in MATLAB 7, in which it is better to use anonymous functions.

```
sol = dsolve('Dy = t^2 + y', 'y(0) = 2', 't');
fsol = inline(vectorize(sol), 't')
```

int Integration operator for both definite and indefinite integrals.

```
int('1/(1 + x^2)', 'x')
syms x; int(exp(-x), x, 0, Inf)
```

inv Inverse of a square matrix.

```
inv([1 2; 3 5])
```

jacobian Computes the Jacobian matrix, *i.e.*, the matrix of partial derivatives with respect to the indicated variables.

```
syms x y; jacobian([x^2*y y-x], [x y])
```

laplace Computes the Laplace Transform.

```
syms t s; laplace(t^5, t, s)
```

length Returns the number of elements in a vector or string.

```
length('abcde')
```

limit Finds a two-sided limit, if it exists. Use **right** or **left** for one-sided limits.

```
syms x; limit(sin(x)/x, x, 0)
syms x; limit(1/x, x, Inf, 'left')
```

load Loads Workspace variables from a disk file with a `.mat` suffix.

```
load filename
```

lookfor Searches for a specified string in the first line of all M-files found in the MATLAB path. Can be slow if many toolboxes are installed.

```
lookfor ode
```

maple Executes a command in the Maple kernel. *Not available in the Student Version.*

```
maple('dsolve', 'diff(y(t),t) + y(t)^2', 'y(t)', 'series')
```

mhelp Gets help about a Maple kernel command or function. *Not available in the Student Version.*

```
mhelp AiryAi
```

more Turns on (or off) page-by-page scrolling of MATLAB output. Use the space bar to advance to the next page, the ENTER or RETURN key to advance line-by-line, and **q** to abort the output.

```
more on
more off
```

num2str Converts a number to a string. Useful in programming.

```
constant = ['a' num2str(1)]
```

ode45 Numerical ODE solver for first order equations. See MATLAB's online help for **ode45** for a list of other MATLAB ODE solvers.

```
[t, y] = ode45(@(t, y) t^2 + y, [0 10], 1);
plot(t, y)
```

odeset Used to set options for **ode45**, such as **Events**, **AbsTol**, and **RelTol**.

```
options = odeset('Events', @eventfile)
```

ones Creates a matrix of ones.

```
ones(3), ones(3,1)
```

path Without an argument, displays the search path. With an argument, sets the search path. Type **help path** for details.

pause Suspends execution of an M-file until the user presses a key.

pretty Displays a symbolic expression in a more readable format.

```
syms x y; expr = x/(x - 3)/(x + 2/y)
pretty(expr)
```

publish Runs the specified M-file and assembles input, comments, and output into a finished document in specified format. (The default is a web page.)

```
publish('mymfile', 'doc')
```

pwd Shows the name of the current (working) directory.

quadl Numerical integrator. The input function must be "vectorized."

```
quadl(@(x) (1-x.^2).^(1/2), -1, 1)
```

quit Terminates a MATLAB session.

roots Finds the roots of a polynomial whose coefficients are given by the elements of the vector argument of **roots**.

```
roots([1 1 -2])
```

save Saves Workspace variables to a specified file. See **load**.

```
save filename
```

sim Runs a Simulink model. Specifying the time interval is optional.

```
sim('mymodel', [0, 5])
```

simplify Simplifies an algebraic expression.

```
syms x; simplify((x+1)^2 - x^2)
```

size Returns the number of rows and the number of columns in a matrix.

```
A = [1 3 2; 4 1 5]
[r, c] = size(A)
```

solve Solves an algebraic equation. Will guess what the independent variable is if it is not specified.

```
solve('x^2 + 3*x + 1')
```

strcat Concatenates two or more strings.

```
strcat('This ', 'is ', 'a ', 'long ', 'string')
```

str2num Converts a string to a number. Useful in programming.

```
constant = 'a7'
index = str2num(constant(2))
```

struct Used to create a "structure."

```
x = struct; x.first = 'a'; x.second = 'b'; x
```

subs Substitutes for parts of an expression.

```
subs('x^3 - 4*x + 1', 'x', 2)
subs('sin(x)^2 + cos(x)', 'sin(x)', 'z')
```

sum Sums a vector, or sums the columns of a matrix.

```
k = 1:10; sum(k)
```

sym Creates a symbolic variable or number.

```
pi = sym('pi')
x = sym('x')
constant = sym('1/2')
```

syms Shortcut for creating symbolic variables. The command **syms x** is equivalent to **x = sym('x')**.

```
syms x y z
```

symsum Evaluates a symbolic sum in closed form.

```
syms x n; symsum(x^n/n, n, 1, inf)
```

taylor Gives a Taylor polynomial approximation of order one less than a specified number (the default is 6) around a specified point (default 0). Will guess what the independent variable is if it is not specified.

```
syms x; taylor(cos(x), 8, 0)
taylor(exp(1/x), 10, x, inf)
```

transpose Transpose of a matrix. Converts a column vector to a row vector, and vice versa. Usually invoked with the **.'** operator. See also **ctranspose**.

```
A = [1 3 4]
A.'
```

type Displays the contents of a specified file.

```
type myfile.m
```

vectorize Vectorizes a symbolic expression. Useful in defining functions; see also **eval** and **inline**.

```
f = inline(vectorize('x^2 - 1/x'))
```

vpa Evaluates an expression to the specified degree of accuracy using variable precision arithmetic.

```
vpa('1/3', 20)
```

whos Lists current information on all the variables in the Workspace.

zeros Creates a matrix of zeros.

zeros(10), zeros(1,10)

Built-in Functions

abs $|x|$.

acos $\arccos x$.

airy Airy functions; **airy(0, x)** and **airy(2, x)** are linearly independent solutions of Airy's equation.

asin $\arcsin x$.

atan $\arctan x$.

atan2 Four quadrant inverse tangent; **atan2(y, x)** is the angle θ in the same quadrant as (x, y) with $\tan \theta = \frac{y}{x}$.

bessel Bessel functions; **besselj(n, x)** and **bessely(n, x)** are linearly independent solutions of Bessel's equation of order n.

cos $\cos x$.

cosh $\cosh x$.

dirac The Dirac δ-"function," representing a unit impulse at $t = 0$.

erf The *error function* $\operatorname{erf}(x) = (2/\sqrt{\pi}) \int_0^x e^{-t^2}\, dt$.

exp e^x.

expm Matrix exponential.

gamma The *gamma function* $\gamma(x) = \int_0^\infty t^{x-1} e^{-t}\, dt$.

heaviside The unit step function with discontinuity at $t = 0$.

hypergeom The Gauss hypergeometric function ${}_2F_1(a, b; c; z)$.

imag The imaginary part of a complex number.

log The natural logarithm $\ln x = \log_e x$.

real The real part of a complex number.

sin $\sin x$.

sinh $\sinh x$.

sinint The *sine integral* $\operatorname{Si}(x) = \int_0^x (\sin t/t)\, dt$.

sqrt $\sqrt{x}$.

tan $\tan x$.

tanh $\tanh x$.

Graphics Commands

axis Sets axis scaling and appearance.

```
axis([xmin xmax ymin ymax])
```
– sets ranges for the axes.

axis tight – sets the axis limits to the full range of the data.

axis equal – makes the horizontal and vertical scales equal.

axis square – makes the axis box square.

axis off – hides the axes and tick marks.

close Closes the current figure window; **close all** closes all figure windows.

contour Plots the level curves of a function of two variables; usually used with the **meshgrid** command.

```
[X, Y] = meshgrid(-3:0.1:3, -3:0.1:3);
contour(X, Y, X.^2 - Y.^2)
```

ezplot Basic plot command for symbolic expressions. Can also be used for implicit plots of $f(x, y) = 0$.

```
ezplot('exp(-x^2)', [-5, 5])
syms x; ezplot(sin(x))
syms x y; ezplot(x^3 - x - y^2, [-2, 2])
```

figure Creates a new figure window.

gca Get current axes. Returns the identifier ("handle") of the **axes** properties of the active figure window.

gcf Get current figure. Returns the number of the active figure window.

ginput Gathers coordinates from a figure using the mouse (press the ENTER or RETURN key to finish).

```
[X, Y] = ginput
```

gtext Places a text label using the mouse.

```
gtext('Region of instability')
```

hold Holds the current graph. Superimpose any new graphics generated by MATLAB on top of the current figure.

```
hold on
hold off
```

legend Creates a legend for a figure.

```
t = (0:0.1:2)*pi;
plot(t, cos(t), t, sin(t))
legend('cos(t)', 'sin(t)')
```

meshgrid Creates a vector array that can be used as input to commands such as **contour** or **quiver**.

```
[X, Y] = meshgrid(0:0.1:1, 0:0.1:2);
contour(X, Y, X.^2 + Y.^2)
```

plot Plots vectors of data. Specifying the plot style is optional.

```
X = [0:0.1:2];
plot(X, X.^3)
plot(X, X.^3), 'rx-', 'LineWidth', 2)
```

plot3 Creates 3-dimensional plots.

```
t = [0:0.1:30];
plot3(t, t.*cos(t), t.*sin(t))
```

print Sends the contents of the current figure window to the printer or to a file. With the **-s** option, prints the current Simulink model.

```
print
print -deps picture.eps
print -s
```

quiver Plots a vector field for a pair of functions of two variables. It takes numerical, rather than symbolic, data. It is often useful to normalize vectors so that they have length close to 1 and to scale them by a factor of about 1/2.

```
[X, Y] = meshgrid(-2:0.2:2, -2:0.2:2);
U = X.*(X - Y - 2); V = Y.*(-X + Y + 1);
L = sqrt(1 + U.^2 + V.^2));
quiver(X, Y, U./L, V./L, 0.5)
```

set Sets a property of a figure window.

```
set(gca, 'FontSize', 14)
```

simplot Similar to **plot**, but uses the style of a Simulink Scope window.

```
t = [0:0.1:10]'; simplot(t, [sin(pi*t), cos(pi*t)])
```

subplot Breaks the figure window into a grid of smaller plots.

```
subplot(2, 2, 1), ezplot('x^2')
subplot(2, 2, 2), ezplot('x^3')
subplot(2, 2, 3), ezplot('x^4')
subplot(2, 2, 4), ezplot('x^5')
```

text Annotates a figure, by placing text at specified coordinates.

```
text(x, y, 'string')
```

title Assigns a title to the current figure window.

```
title 'Nice Picture'
```

xlabel Assigns a label to the horizontal coordinate axis.

```
xlabel 'Year'
```

ylabel Assigns a label to the vertical coordinate axis.

```
ylabel 'Population'
```

zoom Rescales a figure by a specified factor; **zoom** by itself enables use of the mouse for zooming in or out.

```
zoom
zoom(4)
```

Built-in Constants

i $i = \sqrt{-1}$.

Inf ∞. Equivalent to **inf**.

pi π.

MATLAB Programming

end Terminates an **if**, **for**, **while**, or **switch** statement.

else Alternative in a conditional statement. See **if**.

elseif Nested alternative in a conditional statement. See the online help for **if**.

error Displays an error message and aborts execution of an M-file.

for Repeats a block of expressions a specified number of times. Must be terminated by **end**.

```
figure, hold on
t = -1:0.05:1;
for k = 0:10
   plot(t, t.^k)
end
```

function Always used at the top of a function M-file to define a new function.

```
function y = myfunction(x)
```

if Conditional execution of MATLAB statements. Must be terminated by **end**.

```
if (x >= 0)
   sqrt(x)
else
   error('Invalid input')
end
```

input Used in a script M-file to prompt for user input.

```
answer = input('Please enter [x, y] coordinates:  ')
```

isa Used in programs to check whether an object is of a given class (**double**, **sym**, *etc.*).

```
isa(x, 'sym')
```

keyboard Used in an M-file to return control to the keyboard. Useful for debugging M-files.

nargin In M-files, returns the number of arguments passed to the function.

```
if (nargin < 2); error('Wrong number of arguments'); end
```

MATLAB has many other programming constructs, including **`while`**, **`switch`**, **`case`**, **`otherwise`**, **`break`**, **`nargout`**, and **`return`**. Looking at MATLAB's built-in M-files is a good way to learn how to use these.

Simulink Blocks

Algebraic Constraint (Math Operations Library) Equation solver like **`fzero`**.

Clock (Sources Library) Outputs the simulation time.

Constant (Sources Library) Outputs a specified constant.

Demux (Signal Routing Library) Disassembles a vector signal into its components.

From Workspace (Sources Library) Takes input from the MATLAB Workspace.

Gain (Math Operations Library) Multiplies by a constant or a constant matrix.

Integrator (Continuous Library) Computes the definite integral, using a specified initial condition.

Math Function (Math Operations Library) Computes exponentials, logarithms, *etc.*

Mux (Signal Routing Library) Assembles scalar signals into a vector signal.

Polynomial (Math Operations Library) Evaluates a polynomial function.

Product (Math Operations Library) Multiplies or divides signals. Can also invert matrix signals.

Ramp (Sources Library) Outputs a function that is initially constant and then increases linearly starting at the specified time.

Scope (Sinks Library) Plots a signal as a function of time.

Sine Wave (Sources Library) Outputs a sine wave. You can adjust the amplitude, frequency, and phase.

Sum (Math Operations Library) Adds or subtracts inputs.

To Workspace (Sinks Library) Outputs a signal to the Workspace.

Trigonometric Function (Math Operations Library) Computes trigonometric or hyperbolic functions.

Unit Delay (Discrete Library) Useful in modeling difference or differential/difference equations.

Sample Solutions

Here we present complete solutions to selected problems from the Problem Sets. They were prepared using the method discussed in Section 4.4.5. In other words, we prepared a script M-file for each problem, and then "published" the script, in this case to LaTeX. We then assembled the LaTeX files for the published solutions to create this chapter. The only change we made in the output to **`publish`** was to resize the figures and wrap long lines. Note that a sample solution to Problem 2 in Problem Set A already appears in Section 4.4.5.

Problem Set A, Problem 4

Contents

- Clear variables
- Part (a)
- Part (b)

Clear variables

```
clear all
```

Part (a)

We first declare x to be symbolic with syms, and apply factor to the given polynomial. Then we check the answer by expanding it.

```
syms x
factor(x^3 + 5*x^2 - 17*x - 21)
expand(ans)
```

```
ans =

(x-3)*(x+7)*(x+1)
```

```
ans =

x^3+5*x^2-17*x-21
```

Part (b)

We can apply factor directly to the given integer. The output below means that 987654321 = 3^2 * 17^2 * 379721.

```
factor(987654321)
```

```
ans =

           3           3          17          17      379721
```

Problem Set B, Problem 5

Contents

- Clear variables and figures
- Part (a): solving the equation
- Part (a): defining the solution function
- Part (b)
- Part (c): graph of the level curve
- Part (c): graph of the solution
- Part (d)

Clear variables and figures

```
clear all
close all
```

Part (a): solving the equation

As the warning message below indicates, dsolve gives an implicit solution:

```
eqn = 'Dy = (t - exp(-t))/(y + exp(y))';
sol = dsolve(eqn, 't')
```

```
Warning: Explicit solution could not be found; implicit
solution returned.

sol =

1/2*t^2+exp(-t)-1/2*y^2-exp(y)+C1 = 0
```

Part (a): defining the solution function

Since C1 is an arbitrary constant, the general solution is 1/2*t^2 + exp(-t) - 1/2*y^2 - exp(y) = c. Let's define the left-hand side as an anonymous function f.

```
f = @(t, y) (1/2)*t.^2 + exp(-t) - (1/2)*y.^2 - exp(y)
```

```
f =

    @(t, y) (1/2)*t.^2 + exp(-t) - (1/2)*y.^2 - exp(y)
```

Part (b)

Following the prescription for a contour plot, we get the figure below. The final argument to contour makes all the curves black so that they are easier to see when printed. Some of the curves have two y values for each t value. These curves represent two different solutions that meet when their slopes become infinite; this does not violate the existence and uniqueness theorem because they meet only where dy/dt does not exist.

```
[T, Y] = meshgrid(-1:0.05:3, -2:0.05:2);
contour(T, Y, f(T, Y), 30, 'k')
```

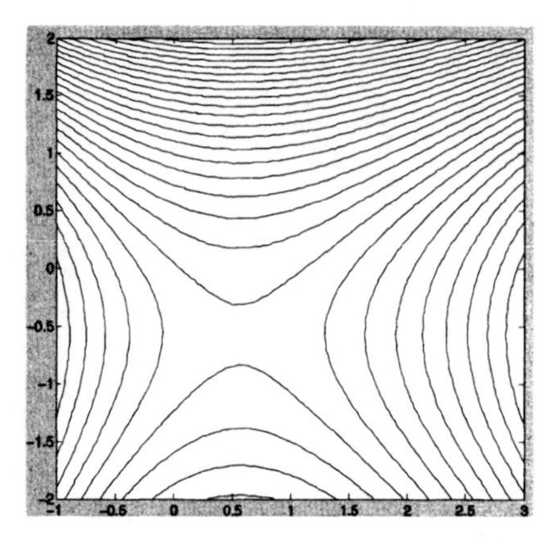

Part (c): graph of the level curves

The solution to the IVP is obtained by setting f(t, y) equal to f(1.5, 0.5). Here we plot it in a separate figure window from the previous graph.

```
c = f(1.5, 0.5);
figure
contour(T, Y, f(T, Y), [c c], 'k')
```

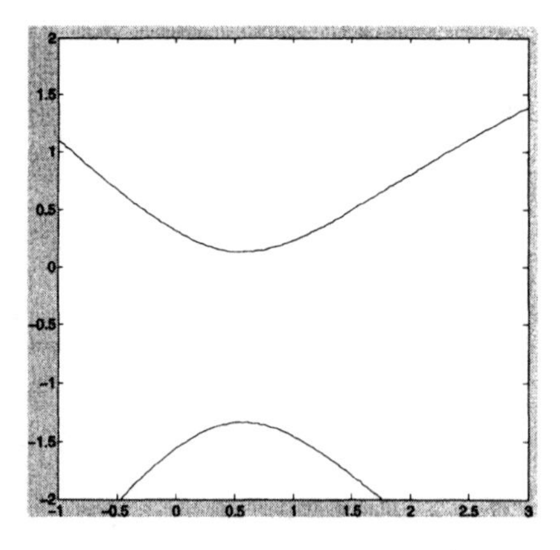

Part (c): graph of the solution

The solution looks like a hyperbola, but only the upper branch satisfies the initial condition. We focus in on this branch by cutting down the range of the axes.

```
axis([-1, 3, -1, 2])
```

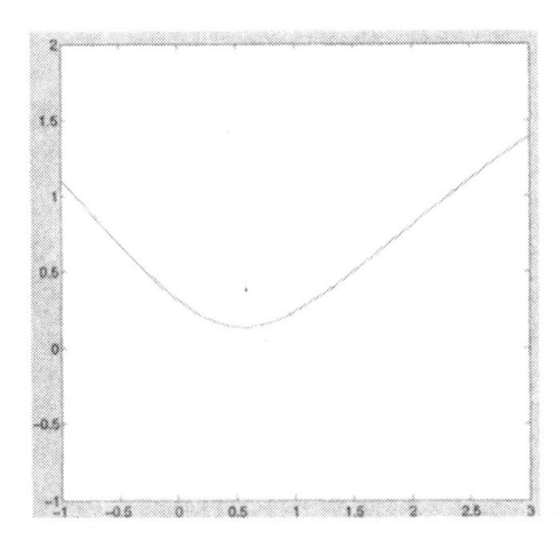

Part (d)

Here are the values of the solution of the IVP at the points t = 0, 1, 1.8, 2.1. At the same time we compute the solution, we mark the points on the graph.

```
hold on
for t = [0, 1, 1.8, 2.1]
    f1 = @(y) f(t, y) - c;
    y1 = fzero(f1, 0.5);
    % Print values of t and y
    [t, y1]
    plot(t, y1, 'o')
end
hold off

ans =

         0    0.3184


ans =

    1.0000    0.2356


ans =

    1.8000    0.6822
```

```
ans =

    2.1000    0.8662
```

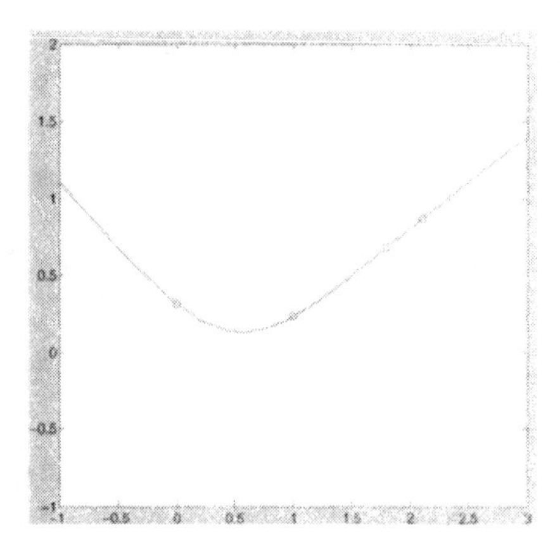

Problem Set C, Problem 3

Contents

- Clear variables and figures
- Part (a): symbolic solution
- Part (a): numerical solution
- Part (a): comments
- Part (b)
- Part (c)
- Part (d)

Clear variables and figures

```
clear all
close all
```

Part (a): symbolic solution

We are going to investigate the IVP y' = (y - t)*(1 - y^3), y(0) = b, for various choices for the initial value b. We first see if we can find the exact (symbolic) solution.

```
dsolve('Dy = (y - t)*(1 - y^3)', 't')
```

```
Warning: Explicit solution could not be found.

ans =

[ empty sym ]
```

Part (a): numerical solution

We see that dsolve is not able to find an exact solution. We thus solve the problem numerically. We define the right-hand side of the differential equation as an anonymous function, and then use a for loop to solve the initial value problem with ode45 for b = -1, -1/2, 0, 1/2, 1 , 3/2, and 2. For brevity we plot all of the numerical solutions on one graph.

```
f = @(t, y) (y - t).*(1 - y.^3)
figure, hold on
for b = -1:0.5:2
   disp(['Solving for initial condition b = ' num2str(b)])
   [t, y] = ode45(f, [0 5], b);
   plot(t, y)
end
hold off
axis([0 5 -5 5])
title 'Solutions for b = -1, -1/2, 0, 1/2, 1, 3/2, 2'
```

```
f =

    @(t, y) (y - t).*(1 - y.^3)

Solving for initial condition b = -1
Warning: Failure at t=2.169418e-01.  Unable to meet
integration tolerances without reducing the step size below
the smallest value allowed (4.440892e-16) at time t.
Solving for initial condition b = -0.5
Warning: Failure at t=5.698501e-01.  Unable to meet
integration tolerances without reducing the step size below
the smallest value allowed (1.776357e-15) at time t.
Solving for initial condition b = 0
Warning: Failure at t=1.188000e+00.  Unable to meet
integration tolerances without reducing the step size below
the smallest value allowed (3.552714e-15) at time t.
```

```
Solving for initial condition b = 0.5
Warning: Failure at t=2.119581e+00.  Unable to meet
integration tolerances without reducing the step size below
the smallest value allowed (7.105427e-15) at time t.
Solving for initial condition b = 1
Solving for initial condition b = 1.5
Solving for initial condition b = 2
```

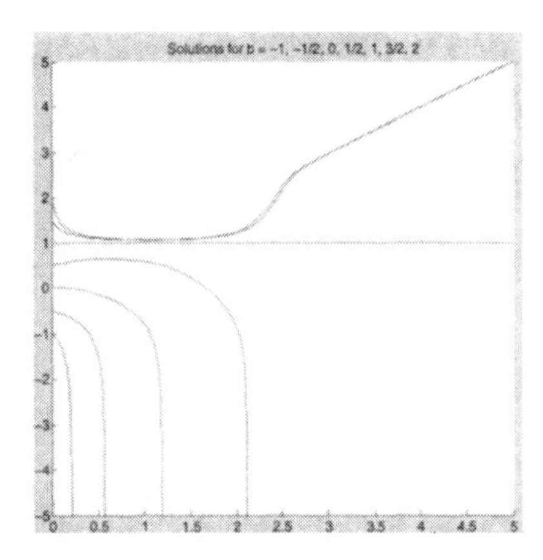

Part (a): comments

For b = -1, -1/2, 0, and 1/2 we get warning messages. For b = -1, the warning and the graph suggest that the solution approaches -infinity as t approaches 0.216... from the left. Likewise, we see that the solutions for b = -1/2, 0, and 1/2 approach -infinity as t approaches 0.569..., 1.18..., and 2.11..., respectively. Note that we displayed the b values so that we could tell which value of b corresponds to which warning message.

Part (b)

The graph in part (a) indicates that if b = 1, the corresponding solution is y = 1; if 0 < b < 1, the solution first increases, and then decreases to -infinity; if b <= 0, the solution decreases to -infinity; and if b > 1, the solution first decreases and then increases slowly to +infinity. That y = 1 is the solution corresponding to b = 1 can be verified by substituting y = 1 into the IVP. It also appears that for values of b less than 1, the solutions have vertical asymptotes at finite values of t. One cannot be certain about this from the graph, but one can in fact show this to be true by a technique similar to that in problems 1 and 2 from this set. As we have noted, it is also suggested by the warning messages produced by ode45.

Part (c)

Now we combine the graph of part (a) with the graph of y = t. It appears that for b > 1, the solutions are asymptotic to the line y = t. To understand this asymptotic behavior, note that along the line y = t, y' is zero; above this line, y' is negative; and below the line y' is positive. Thus the solutions are pushed toward the line y = t.

```
hold on
ezplot('t', [0 5])
title 'Combined Graph'
```

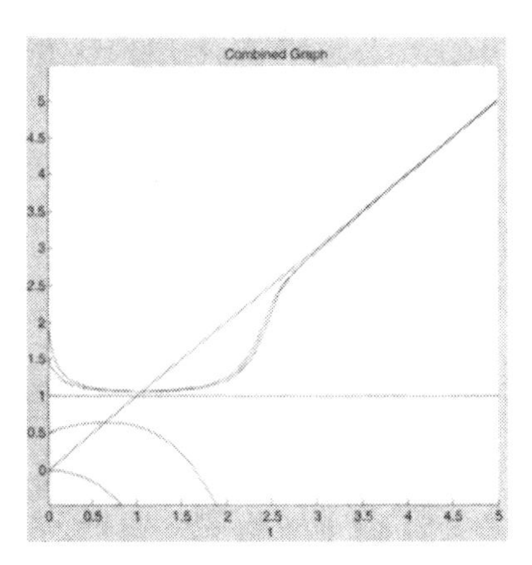

Part (d)

Now we superimpose a plot of the direction field for the differential equation. The result reinforces the observations we have made about the solutions.

```
[T, Y] = meshgrid(0:0.2:5, -2:0.25:5);
S = (Y - T).*(1 - Y.^3);
L = sqrt(1 + S.^2);
quiver(T, Y, 1./L, S./L, 0.5);
title 'Combined Graph with the Direction Field'
```

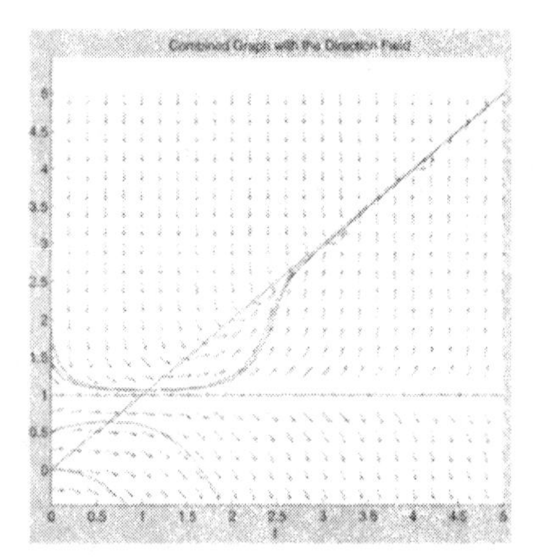

Problem Set D, Problem 1

Contents

Clear variables and figures

```
clear all
close all
```

Part (a)

Here we plot a numerical solution of Airy's equation (as a dashed line) and an exact solution of the facsimile equation (solid line) near t = 0. The two solutions agree well for t between -1 and 1, but they diverge rapidly outside this range.

```
airyeq = @(t, y) [y(2); t*y(1)];
[tfor, yfor] = ode45(airyeq, [0, 2], [0, 1]);
plot(tfor, yfor(:,1), '--')
hold on
[tbak, ybak] = ode45(airyeq, [0, -2], [0, 1]);
plot(tbak, ybak(:,1), '--')
```

```
facA = dsolve('D2y = 0', 'y(0) = 0', 'Dy(0) = 1')
ezplot(facA, [-2, 2])
hold off
title 'Numerical solution and facsimile near t = 0'
```

```
facA =

t
```

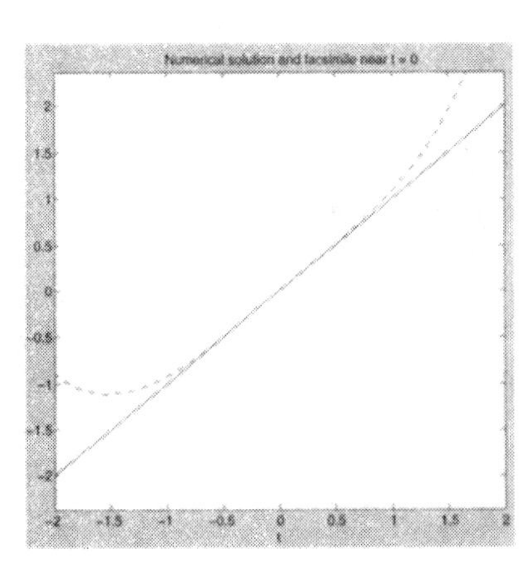

Part (b): numerical solution

Here is a numerical solution to Airy's equation for t near -16 = -4^2. This time we use deval to extract the numerical solution (but not its derivative) at the desired values of t. We also compute the numerical solution to t = -20 so that we can use it again later in part (d).

```
solb = ode45(airyeq, [0, -20], [0, 1]);
tt = -18:0.05:-14;
yy = deval(solb, tt, 1);
plot(tt, yy, '--')
hold on
[min(yy) max(yy)]
```

```
ans =

   -0.6362    0.6472
```

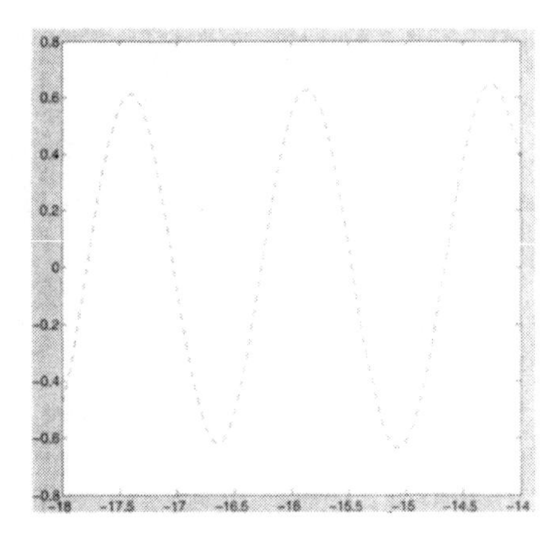

Part (b): facsimile solution

After viewing the graph, we see that it does resemble a sine curve. However, it probably does not lie close to the solution of y" = -16*y with the same initial condition at t = 0 because the two differential equations are not close to each other near t = 0. The amplitude of the oscillation is, from the graph, about 0.6, but we can make this more precise by looking at the data. We printed above the minimum and maximum values of the numerical solution yy between t = -18 and -14, and judge the amplitude to be about 0.64. So our facsimile solution should be y = 0.64*sin(4*t + c2) for some c2. Since the graph crosses zero in the increasing direction near t = -16.25, we choose 4*(-16.25) + c2 = 0, or c2 = 65. The graphs agree quite well over the entire interval.

```
facB = '0.64*sin(4*t + 65)';
ezplot(facB, [-18, -14])
hold off
axis([-18 -14 -1 1])
title 'Numerical solution and facsimile near t = -16'
```

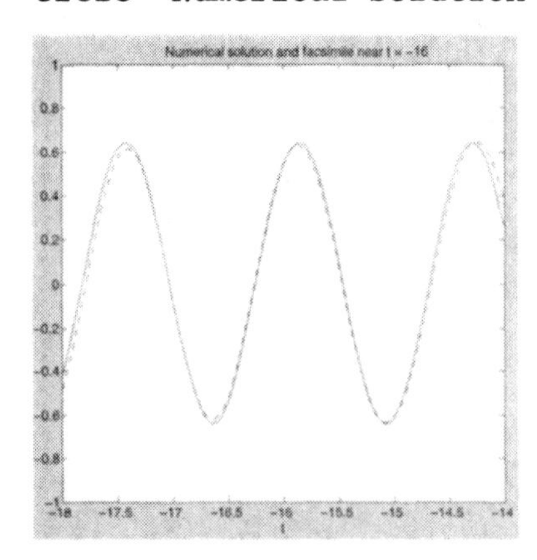

Part (c): numerical solution

Now here is a numerical solution to Airy's equation for t near 16 = 4^2.

```
solf = ode45(airyeq, [0, 18], [0, 1]);
tt = 14:0.05:18;
yy = deval(solf, tt, 1);
plot(tt, yy, '--')
hold on
axis([14 18 0 4*10^21])
```

Part (c): facsimile solution

The solution appears to be growing exponentially, which is also what the hyperbolic sine does far from zero. In fact, sinh(x) = (exp(x) - exp(-x))/2, or approximately exp(x)/2 for large x. So, the proposed facsimile solution is approximately c1*exp(4*t + c2)/2 = (c1*exp(c2)/2)*exp(4*t). This suggests that we can choose c2 arbitrarily (say c2 = 0) and choose c1 so that the magnitudes of the solutions match. Below we choose c1 to make them match at t = 18. Again the graphs agree well, though since the solutions are so much larger between t = 17 and 18 than on the rest of the interval, we can't see how well they agree over the whole interval from a standard graph. In a separate figure, we show the two solutions in a "semilog" graph, where the vertical axis has a logarithmic scale. On such a scale, the graph of an exponential function should be straight. We see that the two solutions do agree well over the entire interval.

```
c1 = 2*yy(end)/exp(4*18)
facC = @(t) c1*sinh(4*t);
ezplot(facC, [14, 18])
hold off
axis([14 18 0 4*10^21])
```

```
title 'Numerical solution and facsimile near t = 16'

figure
semilogy(tt, yy, '--', tt, facC(tt), 'b')
title 'Semilog plot of solution and facsimile near t = 16'

c1 =

    4.2500e-10
```

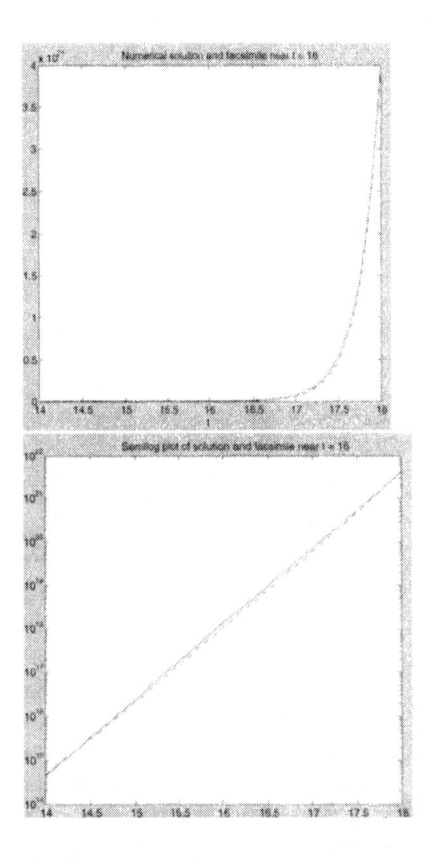

Part (d)

Finally, we show the numerical solution over the interval (-20,2). We use the numerical solutions solb and solf that we computed in parts (b) and (c). From the graph below, it appears that the frequency increases and the amplitude decreases as t goes to minus infinity. The increasing frequency is predicted by the analysis in part (b), where the approximate frequency K increases like sqrt(-t). The decreasing amplitude is not explained by the facsimile approach, at least not without some more detailed analysis.

```
ttb = -20:0.05:0;
yyb = deval(solb, ttb, 1);
ttf = 0:0.05:2;
yyf = deval(solf, ttf, 1);
plot(ttb, yyb, 'b', ttf, yyf, 'b')
axis([-20 2 -2 3])
title 'Oscillations in the solution to Airy''s equation'
```

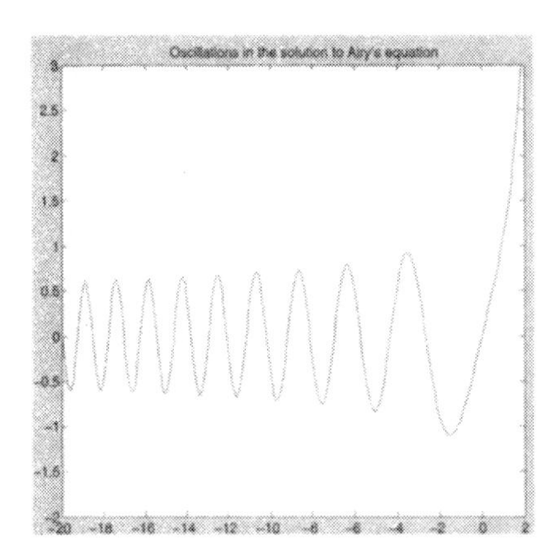

Problem Set E, Problem 1

Contents

Clear variables and figures

```
clear all
close all
```

Part (a): set up

First we compute the formal derivatives that we need, using formalseries.m from Chapter 11.

```
syms x
u = formalseries(x, 10)
du = diff(u, x);
d2u = diff(du, x);
airyser = collect(d2u - x*u, x)

u =

a0+a1*x+a2*x^2+a3*x^3+a4*x^4+a5*x^5+a6*x^6+a7*x^7+a8*x^8+a9*
x^9+a10*x^10



airyser =

-a10*x^11-a9*x^10-a8*x^9+(90*a10-a7)*x^8+(72*a9-a6)*x^7+
(56*a8-a5)*x^6+(42*a7-a4)*x^5+(30*a6-a3)*x^4+(20*a5-a2)*x^3+
(12*a4-a1)*x^2+(6*a3-a0)*x+2*a2
```

Part (a): solution

Here's the series solution up to order 10, using sersol.m from Chapter 11.

```
soln = sersol(airyser, x, 10, [1, 0])

soln =

1+1/6*x^3+1/180*x^6+1/12960*x^9
```

Part (b)

We plot the exact solution as a solid line and the Taylor approximation as a dashed line in each of the graphs below. The approximation looks good between $x = -3$ and 3, though for positive x there is not a clear transition between where the approximation is good and where it isn't. If we restricted the scale on the vertical axis in the graph, the approximation might not look so good for x near 3. The approximation definitely becomes bad for $x < -3$.

```
exact = dsolve('D2y - x*y = 0', 'y(0) = 1', 'Dy(0) = 0', 'x')
ef = @(x) double(subs(exact, 'x', x));
af = @(x) eval(vectorize(soln));

xvals = 0:0.05:5;
plot(xvals, ef(xvals))
hold on
plot(xvals, af(xvals), '--')
title 'Exact and series solutions for positive x'

figure
xvals = -10:0.05:0;
plot(xvals, ef(xvals))
hold on
plot(xvals, af(xvals), '--')
axis([-10 0 -7 1])
title 'Exact and series solutions for negative x'
```

```
exact =

1/2*gamma(2/3)*3^(2/3)*AiryAi(x)+1/2*gamma(2/3)*3^(1/6)*
AiryBi(x)
```

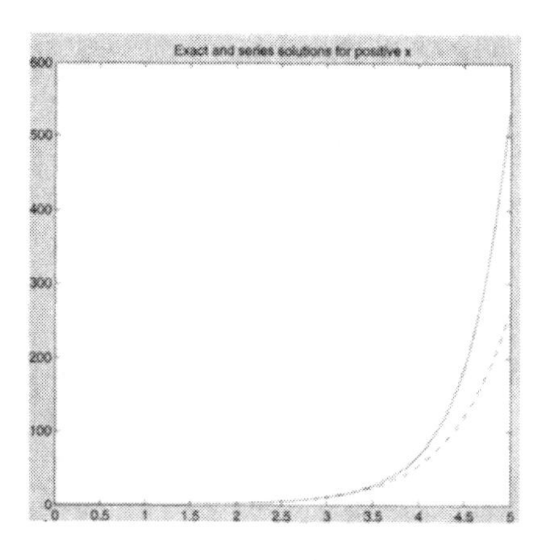

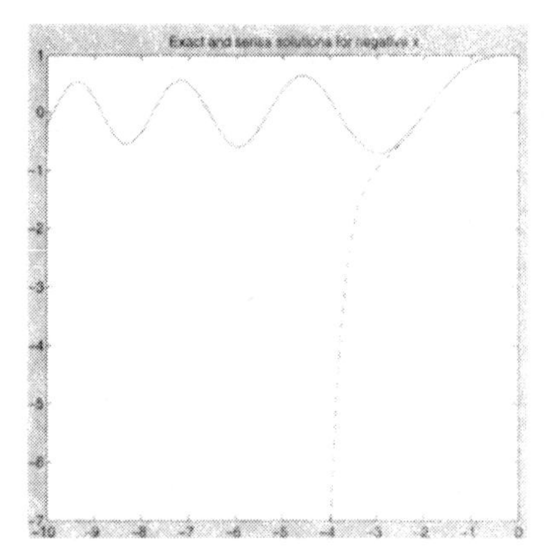

Part (c)

The solutions to y'' = y are linear combinations of the exponential functions y = e^x and y = e^(-x), and any solution that is not a multiple of e^(-x) grows exponentially. The actual solution to y'' = xy with given initial conditions obviously does not decay as x tends to infinity, so it must grow faster than exponentially. The approximate solutions, being polynomials, must grow at a slower rate, and so they must eventually diverge from the actual solution. The first graph in part (b) clearly shows the exact solution growing faster than the approximation as x increases.

Part (d)

The second graph in part (b) indeed shows that the exact solution oscillates for negative x. But the Taylor approximations, being polynomials, must go to plus or minus infinity as x goes to minus infinity, and so must diverge from the exact solution.

Problem Set F, Problem 6

Contents

Clear variables and figures

```
clear all
close all
```

Part (a): critical points

From the output below, we find that the system has four critical points: (0,0), (0,4), (3/2,0), and (1,1).

```
syms x y
sys1 = x*(1.5 - x - 0.5*y);
sys2 = y*(2 - 1.5*x - 0.5*y);
[xc, yc] = solve(sys1, sys2, x, y);
disp('Critical points:'); disp([xc yc])
```

```
Critical points:
[   0,    0]
[ 3/2,    0]
[   0,    4]
[   1,    1]
```

Part (a): stability

First, we compute the Jacobian matrix "A" of partial derivatives of the system and find its eigenvalues "evals" at an arbitrary point (x,y). Then, we substitute the coordinates of each critical point into evals. Based on the results, we classify the critical points as follows:

```
(0,0) -- unstable node;
(0,4) -- asymptotically stable node;
(3/2,0) -- asymptotically stable node;
(1,1) -- unstable saddle point.
```

```
A = jacobian([sys1 sys2], [x y])
evals = eig(A)

disp('Eigenvalues at (0,0):');
disp(double(subs(evals, {x, y}, {0, 0})))
disp('Eigenvalues at (0,4):');
disp(double(subs(evals, {x, y}, {0, 4})))
```

```
disp('Eigenvalues at (3/2,0):');
disp(double(subs(evals, {x, y}, {3/2, 0})))
disp('Eigenvalues at (1,1):');
disp(double(subs(evals, {x, y}, {1, 1})))
```

```
A =

[ 3/2-2*x-1/2*y,          -1/2*x]
[         -3/2*y,      2-3/2*x-y]


evals =

 -7/4*x+7/4-3/4*y+1/4*(x^2+2*x+10*x*y+1-2*y+y^2)^(1/2)
 -7/4*x+7/4-3/4*y-1/4*(x^2+2*x+10*x*y+1-2*y+y^2)^(1/2)


Eigenvalues at (0,0):
    2.0000
    1.5000

Eigenvalues at (0,4):
   -0.5000
   -2.0000

Eigenvalues at (3/2,0):
   -0.2500
   -1.5000

Eigenvalues at (1,1):
    0.1514
   -1.6514
```

Part (b): vector field

Here is a plot of the vector field of the system. The critical points at (0,0) and at (0,4) are easy to recognize, but the other two critical points are less clear. The latter two points, at (3/2,0) and at (1,1), are relatively close to each other, and the direction of the vector field is rapidly changing in this area.

```
[X, Y] = meshgrid(0:0.1:2, 0:0.3:4.5);
U = X.*(1.5 - X - 0.5*Y);
V = Y.*(2 - 1.5*X - 0.5*Y);
L = sqrt((U/2).^2 + (V/4.5).^2);
quiver(X, Y, U./L, V./L, 0.4);
axis tight
xlabel x
ylabel y
title 'Vector Field for Problem 6'
```

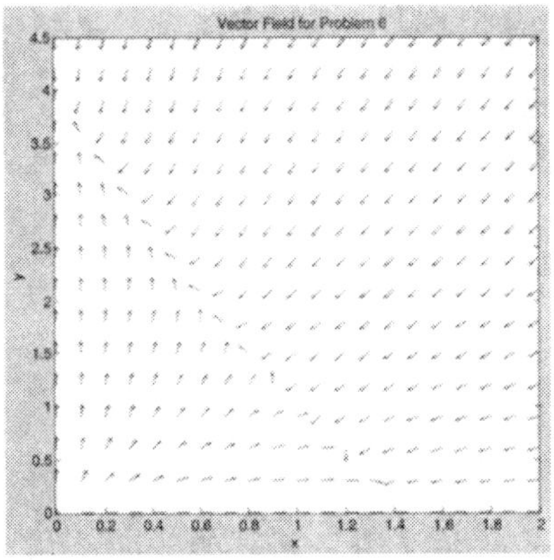

Part (c): first attempt

To plot the phase portrait of the system, we use the approach described at the end of Chapter 13. We first use a rectangular grid to generate different initial conditions. The vector field above makes it clear which direction the solutions are going in the portrait below as t increases.

When following the solutions for negative t, many of them rapidly approach infinity, causing warning messages from ode45. We suppress these warnings with the first command below.

```
warning off all
f = @(t, x) [x(1)*(1.5 - x(1) - 0.5*x(2)); ...
             x(2)*(2 - 1.5*x(1) - 0.5*x(2))];
figure; hold on
for a = 0.25:0.25:1.75
    for b = 0.5:0.5:4
        [t, xa] = ode45(f, [0 10], [a b]);
        plot(xa(:,1), xa(:,2))
        [t, xa] = ode45(f, [0 -5], [a b]);
        plot(xa(:,1), xa(:,2))
```

```
    end
end
axis([0 2 0 4.5])
```

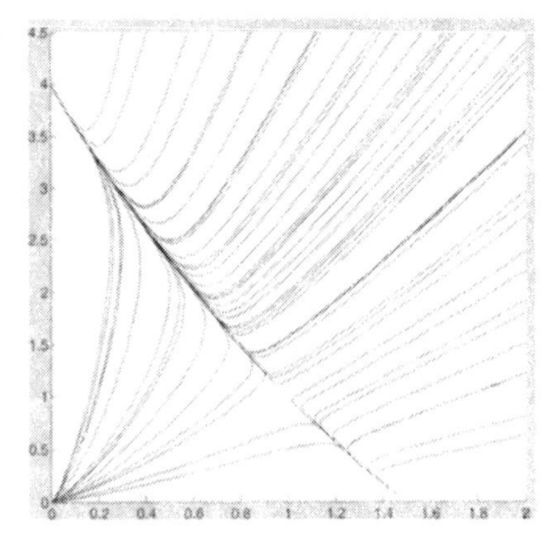

Part (c): improved portrait

The rectangular grid of initial conditions yields unevenly spaced solution curves. Since the solutions rapidly approach a curve connecting the fixed points (1.5,0) and (0,4) and then move along this curve as t increases, we can get a cleaner phase portrait by choosing initial conditions along two parallel lines above and below this curve. Let's use the lines from (1,0) to (0,2.5) and from (2,0) to (0,5).

```
figure; hold on
for a = [1 2]
    for b = 0.1:0.1:0.9
        [t, xa] = ode45(f, [0 20], [a*b a*2.5*(1-b)]);
        plot(xa(:,1), xa(:,2))
        [t, xa] = ode45(f, [0 -5], [a*b a*2.5*(1-b)]);
        plot(xa(:,1), xa(:,2))
    end
end
axis([0 2 0 4.5])
xlabel x
ylabel y
title 'Trajectories for Problem 6'
```

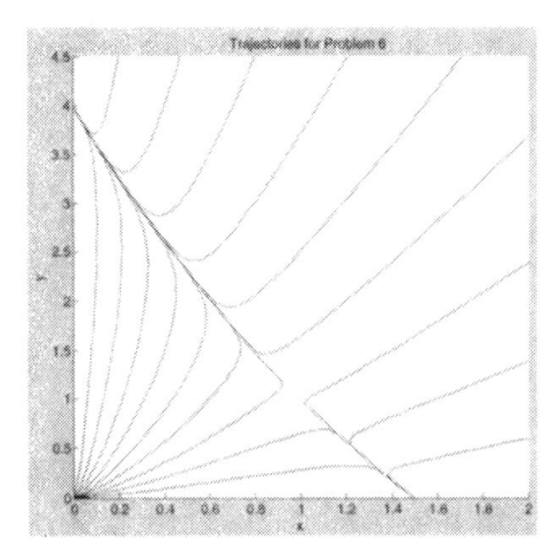

Part (c): combined portrait and vector field

Next, by putting the parametric plot and the vector field plot together we can see the direction of the trajectories. We simply repeat the quiver command from part (b) to redraw the vector field.

```
hold on
quiver(X, Y, U./L, V./L, 0.4)
axis([0 2 0 4.5])
title 'Vector Field and Trajectories for Problem 6'
```

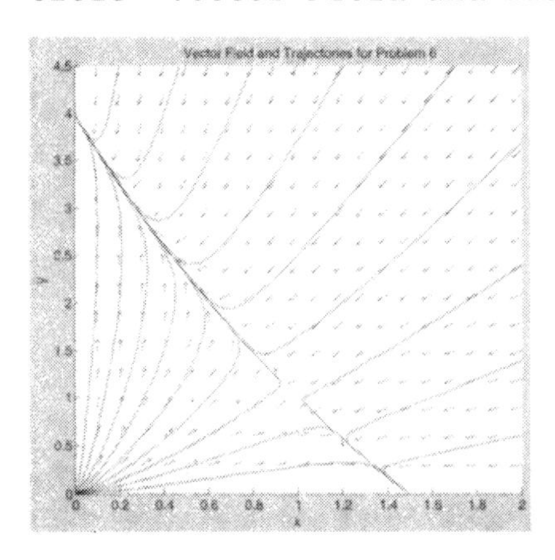

Part (d): table of population values

Here is a table of the populations from times 0 to 20. The first column is t, the second column is x, and the third column is y. The populations change rapidly between t = 0 and t = 1, as they approach the vicinity of the saddle point (1,1), then they change much more slowly. This is consistent with the eigenvalues 0.15 and -1.65 that we computed in part (a) for this equilibrium. The negative eigenvalue has a much larger magnitude than the positive eigenvalue, meaning that in general solutions should approach the saddle point much faster than they leave it.

```
[t, xa] = ode45(f, [0:20], [2.5, 2]);
disp([t xa])
```

```
        0    2.5000    2.0000
   1.0000    1.2945    0.7616
   2.0000    1.2025    0.6299
   3.0000    1.2033    0.5703
   4.0000    1.2219    0.5209
   5.0000    1.2447    0.4722
   6.0000    1.2691    0.4235
   7.0000    1.2934    0.3749
   8.0000    1.3180    0.3279
   9.0000    1.3412    0.2830
  10.0000    1.3638    0.2412
  11.0000    1.3841    0.2030
  12.0000    1.4031    0.1691
  13.0000    1.4193    0.1392
  14.0000    1.4341    0.1136
  15.0000    1.4461    0.0919
  16.0000    1.4567    0.0739
  17.0000    1.4651    0.0590
  18.0000    1.4723    0.0469
  19.0000    1.4779    0.0371
  20.0000    1.4827    0.0293
```

Part (e): discussion and separatrices

There is no peaceful coexistence because almost all the trajectories tend toward an equilibrium point representing a positive population of one species and a zero population of the other. For example, for the initial condition in part (d) we saw that the population of x tends to 1.5 and the population of y tends to 0. In the phase portrait in part (c), some solution curves approach the stable equilibrium point (1.5,0) and some approach the other stable equilibrium point (0,4). The boundary between the initial conditions whose solutions approach (1.5,0) and those whose solutions approach (0,4) is formed by two special solution curves that approach the saddle point (1,1) as t increases. These solution curves are called separatrices.

To approximate the separatrices, we solve the system for negative t using initial conditions near the saddle point (1,1). From the portrait in part (c), we expect one separatrix running from the origin (which is itself an unstable equilibrium point) to (1,1); and another approaching (1,1) from infinity. To find the first separatrix, we choose an initial data point just to the left and below (1,1) and solve backwards in time; to find the second, we do the same thing with an initial data point above and to the right of (1,1). The following commands plot the separatrices as dashed lines, superimposed on the phase portrait from part (c).

```
[t, xa] = ode45(f, [0, -10], [0.99 0.99]);
plot(xa(:,1), xa(:,2), '--');
[t, xa] = ode45(f, [0, -10], [1.01 1.01]);
plot(xa(:,1), xa(:,2), '--');
axis([0 2 0 4.5])
title 'Trajectories for Problem 6, with Separatrices'
```

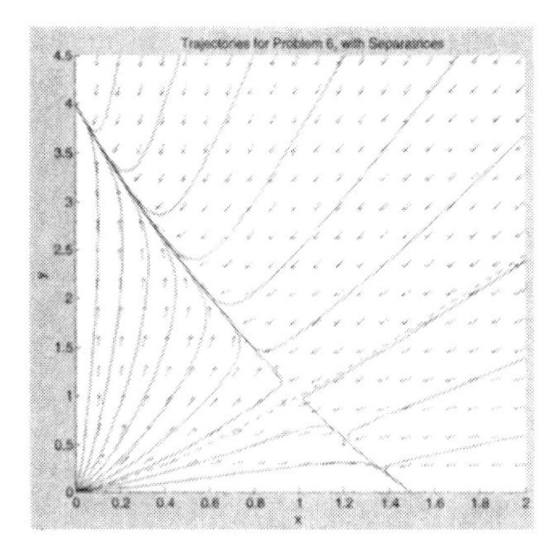

Part (f)

To find a numerical approximation to where the vertical line x = 1.5 cuts the separatrix we delineated above, we repeat the previous ode45 command using an event function that stops execution when the first coordinate of the numerical solution reaches 1.5. We define the event function as an anonymous function, rather than in a separate M-file. To do this, we use the command deal to produce the required three outputs. In addition, because MATLAB's ODE routines sometimes ask the event function for only one output, we need to use the two-argument form of eval, which tries its first argument (which below produces one output), and if that doesn't work, runs its second argument (which below produces three outputs).

We find that the separatrix crosses the line x = 1.5 at approximately y = 1.6898.

```
event = @(t, x) eval('x(1) - 1.5', ...
                'deal(x(1) - 1.5, 1, 1)');
options = odeset('Events', event);
[t, xa] = ode45(f, [0, -10], [1.01 1.01], options);
xa(end,:)
```

```
ans =

    1.5000    1.6898
```

Index

NOTES

NOTES

NOTES

NOTES

NOTES

NOTES

NOTES

NOTES

NOTES

NOTES

NOTES

NOTES

NOTES

CPSIA information can be obtained at www.ICGtesting.com
Printed in the USA
LVOW11s2036150714

394500LV00005B/13/P